A Guide to Radiation Protection

A Guide to Radiation Protection

J. Craig Robertson

Dept. of Physics, Dundee College of Technology, Dundee, Scotland

A Halsted Press Book

JOHN WILEY & SONS
New York

First published 1976 by
The Macmillan Press Ltd
London and Basingstoke

Published in the USA by
Halsted Press, a division of
John Wiley & Sons, Inc., New York

Printed in Great Britain

Library of Congress Cataloging in Publication Data

Craig Robertson, J
  A guide to radiation protection.

  "A Halsted Press book."
  Bibliography: p. 82
  Includes index.
  1. Radioactivity—Safety measures. 2. Radioactive substances—Safety measures.
  I. Title.
  TK9152.C73 1976   614.8'39   76-10458
  ISBN 0-470-18353-5

Contents

Preface

Most of the literature on the subject of radiation protection is written for people who have had some form of scientific training. However, due to the increase in the use of radioactive materials in industry, there are many people who do not have this training but who nevertheless need to understand the basic principles involved in radiation protection. It is the purpose of this book to provide a simple guide for these people. The book is based on experience gained while teaching courses on the subject to firemen, policemen, industrialists and technicians and takes into account their comments on what should be included. They all expressed a wish for a book which would be sufficiently detailed to enable them to handle radioactive materials safely but would not 'blind them with science'.

To make the book easier to use there is a summary at the end of each chapter outlining the important things to remember from that chapter. These summaries form a quick guide to the contents of the book. Each chapter is illustrated where necessary by diagrams or photographs.

The book could also usefully be used as an introduction to the subject of radioactivity at an elementary level.

Acknowledgements

Thanks are due to Dundee College of Technology for allowing the author to use their facilities in making many of the illustrations used in the book. Several colleagues read the draft manuscript of the book and their comments have been taken into account in preparing the final version.

Dundee, 1976   J.C.R.

1

The biological effects of radiation

Time spent in improving one's knowledge of radioactivity and radiation from the point of view of radiation protection is not time wasted. Perhaps the closest that the United Nations Scientific Committee on the Effects of Atomic Radiation comes to making a pronouncement is to agree that the exposure of people to this radiation ought to be kept to a minimum. It is known that if enough radiation is absorbed in the body, it may produce clinical effects in the person who is exposed to it. This is referred to as the *somatic* effect of radiation. Radiation may also produce a *genetic* effect where the effect of the radiation is only apparent in the offspring of the persons exposed to the radiation. That is, the effect of the radiation appears in their children or their children's children and so on. It is for this latter reason that exposure of the public to radiation is carefully controlled.

The natural background radiation

Before considering the somatic and genetic effects of radiation in more detail, it is useful to consider the natural background radiation to which we are all exposed. Part of this background comes from the radioactive materials which are present in our surroundings. When the substances which are found on earth were formed millions of years ago, many of them were highly radioactive. With the passage of time most of them are now inactive, but there are still some active today. For example, uranium and its associated materials, which are present in most salts and rocks, emit radiation which forms a part of the natural background radiation. Thus the granite paving stones found in many areas can result in background radiation which may be several

1

times higher than in other areas, since they contain a high content of uranium. Even the potassium in our bones is radioactive. The earth is also exposed to cosmic radiation which, as the name implies, falls on the earth from outer space and the sun. Most of the cosmic radiation is absorbed in the upper atmosphere but some penetrates to ground level and forms another part of our natural background. Nowadays man has increased the radiation to which he is exposed by the use of man-made radioactive materials in industry, by fallout from the testing of nuclear weapons, by viewing television and by the use of x-rays in medicine and dentistry.

In comparing quantities of radiation we talk of the radiation dose, and the dose received from background sources by each individual during the normal reproductive years can be calculated. It is expressed in a unit called the rem, which will be explained in a later chapter. For the present, in order to establish a means of comparison, it is sufficient to remember that the average person will receive a dose of approximately 3 rems, from background sources, during the 30-year reproductive period of his life. It is comforting to note that in many areas, such as parts of India, the background dose is very much higher than this without apparently producing any ill-effects in the population living there.

Using table 1.1, which is adapted from an American Nuclear Society publication, readers can compute the likely radiation dose they will receive during their lifetime.

The somatic and genetic effects of radiation

Let us now consider the somatic and genetic effects of radiation in more detail. All living organisms are composed of cells and the effect of radiation is to cause damage to these cells. If the damage is to the cells which are concerned with the metabolism of the body, then *somatic* damage to the individual may result. If, however, the damage is to reproductive or potentially reproductive cells, then a *genetic* effect may be the outcome.

The somatic effects

The somatic effects of radiation on the individual can be divided into two classes—early and late. Very large doses of radiation delivered in a short time to a large fraction of the body produce ob-

Table 1.1 Compute your own radiation dose

We live in a radioactive world. Radiation is all about us and is part of our natural environment. By consulting this table, you will get an idea of the amount of radiation you are exposed to every year

Lifestyle	Source of radiation	Annual dose (mrem)
Where you live	Cosmic radiation at sea level; add 1 for every 30 m (100 ft) of altitude	30
	House construction	
	Wood	30
	Concrete	34
	Brick	34
	Stone	36
	Ground	15
What you eat, breathe and drink	Water, food and air	25
How you live	Jet aircraft flights	Number of long flights × 4
	Television viewing	
	Black and white	1 for each hour viewed
	Colour	2 for each hour viewed
	X-ray diagnosis and treatment	150 for each chest x-ray taken
		20 for each dental x-ray taken

Note: 1 mrem $-\frac{1}{1000}$ of a rem.

vious clinical effects within hours, days or weeks. Examples of effects which can occur are loss of weight, changes in the blood cells, loss of hair, reddening or blistering of the skin, nausea and vomiting, changes in fertility and in extreme cases even death. The radiation dose required to produce these effects is very many times the background dose of 3 rems, so fortunately is usually outside normal human experience. The late effects of radiation are various forms of cancer, cataract and sterility. These may occur in persons who have been exposed to very high doses of radiation but may also arise from chronic exposure to lower doses of radiation as experienced, for example, by radiation workers. Our knowledge of these delayed effects come from past experience—from the survivors of the atomic bombing of

Hiroshima and Nagasaki, from children exposed prenatally to x-rays, from children treated for enlarged thymus glands by irradiation and from adults treated by x-rays for chronic arthritic conditions of the spine. Other sources of information include adults treated for thyroid conditions using radioactive iodine, uranium miners exposed to radioactive gases and dust, and individuals who have ingested radium.

All the available information leads to the conclusion that very few people will in fact develop any of the effects of radiation, but that the chance of doing so increases as the radiation dose increases. This is rather similar to lung cancer. Not everyone who smokes will develop it, but the risk increases with the number of cigarettes smoked.

The genetic effects

These arise from changes in the male and female reproductive cells. The body of an adult contains nearly 20 times as many cells as the body of an infant at birth. This increase in the number of cells results from the process of cell division in which the cell produces an exact replica of itself. During this process a definite number of threadlike objects called chromosomes appear and these carry large numbers of genes arranged along the length of the chromosomes in a specific order. The genes determine the hereditary nature of the individual. Damage to the chromosomes or genes results in the emergence in the offspring of characteristics not present in the parents. This is referred to as a mutation. Not all mutations are harmful but a lot are. The natural rate of mutations in all births is about one per cent and while there is little evidence to link radiation with this factor, radiation is known to produce effects in mammals, so the effect of irradiating the population as a whole is likely to be to increase the mutation rate. Hence there could be an increase in the number of malformations and diseases in children.

Maximum permissible dose

Even though there is no direct evidence to suggest that doses of radiation which are about the same as the background radiation are harmful, it is obviously desirable to keep the exposure to radiation of the population as a whole as small as possible. This could most readily be done by stopping the use of all radioactive materials in industry, banning the medical uses of radiation and stopping any further development of nuclear power stations. This seems too high a price to pay, especially when consideration is given to the studies of the effects of radiation. These studies have been much more extensive than similar investigations of any other environmental agent and have shown that radiation produces no unique effects in man. Similar effects can be produced by other agents used by man; for example, certain pesticides and other common chemicals are also known to cause mutations. So if we stopped using radioactive materials, logically we should stop using many other things as well. Nevertheless there should be good justification for the use of radiation in any particular way.

The International Commission on Radiological Protection (I.C.R.P.) points out that the problem in practice is to limit the dose to levels which involve a risk not unacceptable to the individual or the population at large. This is called the *maximum permissible dose* and is based on careful consideration of all the presently available knowledge of the effects of radiation. It is stressed that as more information is gathered the recommended maximum permissible dose may alter.

The I.C.R.P. has recommended maximum permissible dose levels for

    (a) people who are exposed to radiation in the course of their work, and

    (b) members of the public.

Table 1.2 Maximum permissible doses for adult workers aged 18 years and above

Organ	Yearly limit	Three-monthly limit
Gonads and red bone marrow (and, in the case of uniform irradiation, the whole body)	5 rems	3 rems
Skin, thyroid and bone	30 rems	15 rems
Hands, forearms, feet and ankles	75 rems	40 rems
All other organs	15 rems	8 rems

The maximum permissible doses for exposure of workers aged 18 years and above is given in table 1.2. Recommendations are made both for whole body irradiation and for irradiation of limited regions of the body such as the hands. The object of the three-monthly dose limits is to limit to some extent the rate of accumulation of the dose.

Women should be subject to further dose limits. The dose to the abdomen should be limited to $1 \cdot 3$ rems/qu. y and 5 rems/y. If pregnancy has been diagnosed the dose during the remainder of the pregnancy should be limited to 1 rem. Special attention should be given to the type of work done by employees known to be pregnant, to ensure that they are excluded from areas where moderate to high radiation doses might occur as the result of accidents.

For individuals of less than 18 years of age but greater than 16 years of age, the annual dose to the gonads, blood-forming organs or whole body should be limited to $1 \cdot 5$ rems/y.

For members of the public the dose limits are one-tenth those of radiation workers but with a rider that the dose to the thyroid of children up to 16 years of age should not exceed $1 \cdot 5$ rems/y.

On the rare occasions where it may be appropriate to authorize a special exposure, such as an accident involving radioactive materials, the dose authorized should not exceed twice the annual dose limits of table 1.2 in any one event and five times these limits in a lifetime. For whole body irradiation the yearly maximum of 5 rems is derived from a more general relationship that the dose accumulation at any age should not exceed $5 \times (age - 18)$. For example, by age 30 the dose should not exceed 60 rems. In permitting special exposures this rule must also be obeyed.

Derived working limit

To ensure compliance with the I.C.R.P. dose recommendations, *derived working limits* are often used. For example, in the United Kingdom a derived working limit of $2 \cdot 5$ mrem/h is used as the maximum permissible rate of exposure for radiation workers based on a 40-hour week, 50-week year, since this ensures compliance with the dose limits of 5 rems/y ($0 \cdot 0025 \times 50 \times 40 = 5$ rem). Other derived working limits are set in terms of the quantity of radioactive material which is permitted on surfaces, clothing, or the skin and many of the derived working limits used in the protection of the public are specific to the particular circumstances to which they apply. Thus the quantity of strontium, iodine and caesium permitted in milk is carefully controlled.

It is the responsibility of those working with radiation to keep the dose received by themselves and other individuals to an absolute minimum and certainly to within the I.C.R.P. recommendations. The practice whereby dose levels are not reduced to the absolute minimum just because the annual dose will be less than the permitted maximum is to be avoided.

The maximum permissible levels given above are for sources of radiation which are external to the body but there are similar recommendations which apply when the source of radiation is inside the body. Control is effected in terms of the maximum permissible body burden of any material, that is, in terms of the quantity of any material which can safely be accumulated in the body.

Clearly, before it is possible to ensure that the various safety recommendations are adhered to, it is necessary to understand something about radioactivity and the units in which doses are measured. To be able to reduce the dose to the required levels it is necessary to know how to shield radiation and so on. The following chapters in this book are intended to cover all the relevant topics so that the reader is made aware of the steps which should be taken to protect himself and others.

Things to remember from chapter 1

(1) In comparing quantities of radiation, we talk of the radiation dose.
(2) Large doses of radiation can be harmful, producing both somatic and genetic effects.
(3) The somatic effects are produced in the persons exposed to the radiation, while the genetic effects are produced in their children.
(4) Because of the harmful effects of radiation, maximum permissible doses have been set for the amount of radiation which can be received in a year by the various organs in the body and for whole body irradiation. Radiation workers can receive higher doses than the general public. In practice, derived working limits are also used.

(5) It is the responsibility of those working with radiation to keep the quantity of radiation absorbed by themselves and other individuals to an absolute minimum and certainly within the limits recommended by the International Commission on Radiological Protection.

2

What is radioactivity?

The structure of matter

> What are little boys made of?
> Snips and snails and puppy dogs' tails
> That's what little boys are made of.

Well, of course, we all know that the nursery rhyme is not true. But what are things made of? The answer to this question was given a long time ago by chemists who found that all the different types of substances we encounter in our everyday life were combinations of a relatively few number of basic chemical materials called the elements. Salt, for example, is formed by a combination of the elements sodium

Figure 2.1 Illustrating some common elements. This figure is reproduced by courtesy of the Central Electricity Generating Board

and chlorine, water from the elements hydrogen and oxygen. Other elements are shown in figure 2.1. Clearly we can now ask, 'What are the elements made of?' The answer is atoms. The atoms form the smallest part of the elements and for a long time were thought to be indivisible, but we now know that the atoms themselves have a structure and it is changes in this structure which give rise to radioactivity. At the centre of each atom is the nucleus, and in orbit round the nucleus in a similar manner to the earth round the sun, are the electrons. Each element has a different number of electrons in orbit round the nucleus and it is this fact which results in their different appearance and chemical behaviour. Further scientific research has shown that the nucleus itself has a structure and is composed of neutrons and protons. The combination of protons and neutrons in the nucleus and the electrons in orbit round the nucleus make up the atoms. The protons and neutrons are very much heavier than the electrons so the number of protons and neutrons in the nucleus determines how heavy each atom is.

Mass numbers and atomic numbers

Consider a piece of aluminium. Aluminium is an element. Along with all the other chemical elements it is given a chemical symbol, in this case Al. Gold is also an element with the chemical symbol Au. We can now ask whether one atom of gold is heavier than one atom of aluminium. If it is, why is this so? The answer is yes, and the reason is that the nucleus of the gold atom contains more protons and neutrons than the aluminium atom. This difference is expressed by assigning to each a *mass number*. The mass number of an atom tells us how heavy it is in relation to any other atom. Since, as we have said, the mass of an atom is concentrated in the nucleus it also tells us the total number of neutrons and protons in the nucleus. Aluminium has a mass number of 27 which tells us there are 27 protons plus neutrons in the nucleus. For gold the mass number is 197 so that the gold nucleus contains 197 protons plus neutrons and in consequence one atom of gold is heavier than one atom of aluminium. But how many neutrons are there and how many protons? To answer this question we must first consider why the electrons remain in orbit round the nucleus. This is because there is an interaction or force between the protons in the nucleus and the electrons. As an analogy you might imagine that the electrons like to hold hands with the protons but not with the neutrons. There has to be the same number of protons in the nucleus as electrons in orbit round the nucleus. The number of electrons accounts for the chemical behaviour of the elements and this number is called the *atomic number* and a different atomic number is ascribed to each atom. The atomic number of aluminium is 13. There are 13 electrons in orbit round the nucleus and since the number of protons is equal to the number of electrons, there are 13 protons in the nucleus of the aluminium atom. Since the mass number of aluminium is 27 which means there are 27 neutrons plus protons in the nucleus, there are clearly 14 neutrons in the aluminium nucleus. The number of neutrons is denoted by the neutron number N.

The complete classification of any atom is given by $^A_Z X$ where A is the mass number, Z is the atomic number and X stands for the chemical symbol of the element. Thus for aluminium we write $^{27}_{13}Al$. Suppose now we add an extra neutron to the aluminium nucleus so that we now have 15 neutrons and 13 protons in the nucleus. The mass number becomes 28 but the atomic number which tells us the number of electrons in orbit round the nucleus remains unaltered and the atom as a whole still behaves chemically like aluminium. The two different kinds of aluminium atom are referred to as isotopes of aluminium. The atoms have the same chemical behaviour but one is heavier than the other. We can find other atoms with the same mass number as $^{28}_{13}Al$ but not with the same number of protons and neutrons. That is, not with the same Z/N ratio. For example, $^{28}_{14}Si$ and $^{28}_{15}P$ both have the mass number 28, but the silicon, $^{28}_{14}Si$, nucleus contains 14 neutrons and 14 protons, whereas the phosphorus, $^{28}_{15}P$, nucleus contains 15 protons and 13 neutrons. $^{28}_{13}Al$ on the other hand has 15 neutrons and 13 protons in the nucleus. Now for every mass number there are only certain Z/N ratios for which the nucleus is stable or non-radioactive. The other atoms with the same mass number change or 'decay' to form the stable atom. For the mass number 28 it is the silicon nucleus which is stable. The atoms $^{28}_{13}Al$ and $^{28}_{15}P$ are radioactive and change or decay into $^{28}_{14}Si$.

Thus radioactivity is the change of one type of atom into another with the emission of radiation. Later we shall consider the different types of radiation which are emitted.

When we are dealing with phenomena which depend on the composition of the nucleus we use the term *nuclide* rather than atom and this term will be used from now on in this book. Any nuclide is characterized by the number of neutrons and protons within its nucleus.

Quantity of radioactivity

When you visit the supermarket you make use of quantitative units to describe the weight of the goods you buy. In the same way, if you were able to visit a supermarket selling radioactive materials, it would be necessary to have units to determine the amount of radioactive material you were buying and just how much radioactivity you were getting. The unit used to report the activity of any sample of radio-active material is the *curie*. Since radioactivity is the change of one kind of nuclide into another with the emission of radiation, the activity of any radioactive material will be determined by the number of nuclides decaying or changing within it per unit time. A disintegration rate of 37 000 000 000 disintegrations per second which is abbreviated as $3 \cdot 7 \times 10^{10}$ d/s is defined to be 1 curie. The curie is a large unit, and more commonly quantities of material are encountered in which the disintegration rate is one millicurie or one microcurie. These are one-thousandth and one-millionth of a curie respectively. The picocurie, one million-millionth of a curie is also used.

1 curie	37 000 000 000 d/s
1 millicurie	37 000 000 d/s
1 microcurie	37 000 d/s
1 picocurie	0·037 d/s

Curie is abbreviated as Ci, millicurie as mCi, microcurie as μCi and picocurie as pCi.

The curie was originally chosen in relation to the activity of 1 g of radium; 1 g of natural uranium has an activity of approximately 1 millicurie. For some radioactive materials, very little material indeed has an activity of 1 curie. For other materials, a large amount of material is required. This depends on the radioactive nuclide with which we are dealing; this is discussed in chapter 5.

Recently the General Conference on Weights and Measures approved a unit which will eventually replace the curie. This unit is the *becquerel,* abbreviated as Bq. An activity of 1 d/s is defined to be 1 *becquerel.* Hence 1 *becquerel* is equal to approximately 27 picocuries (pCi).

Labelling of radioactive materials

It is important to remember how nuclides are classified and the abbreviations used for the units of radioactivity. All radioactive material is supplied in containers which are labelled using the classification. Very often a delivery note will accompany the material. Thus a 3 curie, ^{60}Co or 60-Co source might be supplied for use in radiography. This is the shorthand for a 3 curie cobalt source in which the active nuclide has a mass number of 60. In addition when radioactive material is stored it should be kept in a container clearly labelled in the same way.

Things to remember from chapter 2

(1) Radioactivity is the spontaneous change of one kind of nuclide (atom) into another with the emission of radiation.
(2) There are several different kinds of radiation which can be emitted.
(3) The activity of any radioactive material is expressed in curies although the curie will eventually be replaced by the *becquerel*.
(4) Curie is abbreviated to Ci, millicurie to mCi and microcurie to μCi; *becquerel* is abbreviated to Bq.
(5) All radioactive material will be supplied with a delivery note specifying the active nuclide and the activity of the material.

3

Recognition and classification of radioactive materials

There is no simple way of telling if a piece of material is radioactive or not. You can't smell radioactivity. This is fairly obvious but is probably worth emphasizing since radioactive material is very often shown to glow, in films and on television for example. In practice this just does not happen. A piece of radioactive aluminium will behave in exactly the same way as a piece of non-radioactive aluminium, except of course that it will emit radiation. Nor is there any relationship between the physical size of a piece of radioactive material and its activity. A small piece of material can be highly active, a large piece of material on the other hand may only be slightly radioactive or indeed inactive. This must be so since radioactive materials are made from inactive materials. In even a minute quantity of material there are so many nuclides that it is always possible to increase the number of active nuclides and hence to increase the activity.

Radioactive material is a source of radiation and hence we speak of radioactive sources. These are classified according to the physical form of the material which shapes the source into *open* and *sealed* sources. Extra precautions have to be taken when handling open sources.

Open and sealed sources

By definition a *sealed source* is any radioactive material sealed in a container or bonded wholly within material which is not itself radioactive. The container and bonding are treated as part of the source. Thus

14

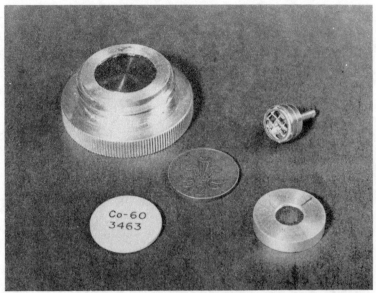

Figure 3.1(a)   Some typical sealed sources

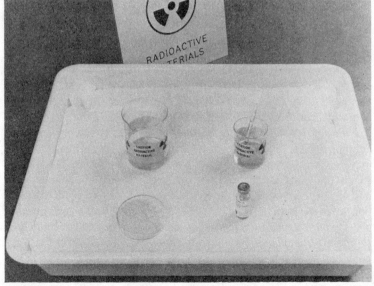

Figure 3.1(b)   Some typical open sources. The material forming these sources can easily be spread about. Since the source material can easily be absorbed into the body such sources are a potential internal radiation hazard

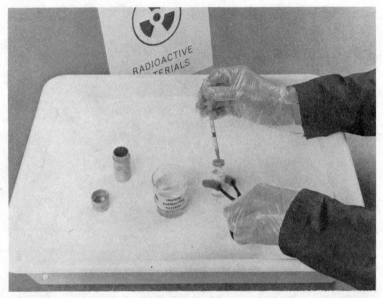

Figure 3.2   Dispensing open source material into a beaker. Note that the operation is carried out over a tray lined with absorbent paper. Gloves and protective clothing are worn whilst handling the radioactive material

a sealed source is in solid form and as a result the radioactive material forming it is localized and unlikely to be spread about. A sealed source is most unlikely to be absorbed within the body so any radiation emitted by the source will present an external radiation hazard and steps can be taken to reduce this hazard.

An *open source* is not a sealed source—this is the legal way of defining open sources. They may be in liquid or gaseous form, or they may be formed by solid materials which are in a powdery form. Open source materials can therefore be very easily spread about and consequently are much more likely to cause *contamination*. By contamination we mean material which has been accidentally spilled or released and which is adhering to surfaces, clothing, floors and so on. Open source material is in a form where it can readily be absorbed into the body and so it is much more hazardous than the material in a sealed source. It forms a potential internal radiation hazard. Once it is in the body it will continue to irradiate the body until either the radioactivity has ceased, which may take years, or until it is excreted by the body. This may only take a few days, but again it may not happen at all.

Examples of open and sealed sources are shown in figures 3.1(*a*) and (*b*). Figure 3.2 shows open source material being dispensed from the ampoule in which it was supplied into a beaker.

Open and sealed sources are usually stored in separate locations.

Radiotoxicity

The International Commission of Radiological Protection has considered the hazards from internal sources and has classified the radionuclides which could form such sources according to their *radiotoxicity*. This classification depends on the type of radiation the nuclide emits, how long the material forming the source will remain radioactive, how rapidly the nuclide is eliminated from the body and whether the nuclide tends to gather in a specific organ within the body. This classification is given in the Appendix which can also be used as a reference for identifying any nuclide by its chemical symbol, mass number and common name. Before open sources are handled a laboratory must be prepared which takes into account the radiotoxicity of the nuclide to be handled, the level of radioactivity to be handled and the type of operation to be performed. The laboratory requirements are given in table 3.1. A low-level laboratory would be of the standard of a good chemical laboratory, a medium-level laboratory would be a special room used only for radioactivity work. A high-level laboratory would be of the kind found in places like the Radiochemical Laboratories at Amersham in England. Highly specialized equipment is provided for handling the active materials prepared and dispensed in these laboratories. The quantity of radioactivity which can be handled in the different types of laboratory can be modified according to the type of operation as indicated in table 3.2.

Table 3.1 Classification of laboratories on the basis of radiotoxicity and quantity of the nuclide to be handled

Radiotoxicity of nuclide	Type of laboratory or working place required		
	Low-level laboratory	Medium-level laboratory	High-level laboratory
Class I high	10 μCi or less	10 μCi–10 mCi	10 mCi or more
Class II medium	100 μCi or less	100 μCi–100 mCi	100 mCi or more
Class III medium	1 mCi or less	1 mCi–1 Ci	1 Ci or more
Class IV low	10 mCi or less	10 mCi–10 Ci	10 Ci or more

Table 3.2   Modifying factors to be applied to table 3.2 according to the
types of procedure to be used

Procedure	Modifying factor
Storage (stock solutions)	×100
Very simple wet operations	×10
Normal chemical operations	×1
Complex wet operations with risk of spill ⎫ Simple dry operations ⎬	×0·1
Dry and dusty operations	×0·01

Suppose for example it became necessary to handle an open source of phosphorus. In the radiotoxicity tables (see the Appendix), phosphorus appears in Class III of medium toxicity. From tables 3.1 and 3.2 it can be seen that 100 mCi could be stored and simple wet operations performed on 10 mCi, in a low-level laboratory. For chemical operations not more than 1 mCi would be handled in a normal chemical laboratory. For a polonium source of a high toxicity, Class I, these quantities would be very much reduced in each case. The radiotoxicity tables can also be used to classify if a particular open source is hazardous or not. Thus a 10 μCi source containing a very highly toxic nuclide would be handled with greater precautions than a 10 mCi source containing a slightly toxic nuclide. Since it would, in general, be very difficult to absorb a sealed source, the radiotoxicity tables are not generally referred to when handling such sources.

Protective clothing is always worn when handling open source materials. Just how elaborate this protective clothing has to be depends on circumstances. For example, when dealing with surface contamination of a slightly toxic nuclide where low activities are involved, a laboratory coat, gloves and overshoes would suffice. In other instances, it might be necessary to use special suits, with a filter mask, or even a mask fitted with its own air supply. For liquid contamination p.v.c. suits are often used.

It is important to notice that just because open sources present a special hazard when absorbed into the body this does not mean they will not present an external hazard in the same way as a sealed source. Both open and sealed sources emit radiation which when it falls on the body may result in damage, as discussed in chapter 1.

As a rule of thumb when we are considering both open and sealed sources external to the body: *curie sources will in general be very hazardous; millicurie sources will in general be less hazardous; microcurie sources will in general be fairly safe.*

Things to remember from chapter 3

(1) It is not possible to tell by appearance whether a substance is radioactive or not.
(2) Radioactive materials are subdivided into two classes: open and sealed sources.
(3) Open sources are handled with even greater respect than sealed sources because of the danger of absorption into the body.
(4) The radiotoxicity tables given in the Appendix can be used along with tables 3.1 and 3.2 to classify how hazardous a particular open source is.
(5) For sealed sources: curie sources will in general be very hazardous, millicurie sources less hazardous, and microcurie sources fairly safe.

4

Types of radioactive source

In chapter 2, radioactivity was defined as the spontaneous change of one nuclide into another with the emission of radiation. The nuclides are said to be radioactive and the transformation process is referred to as radioactive decay. Radioactive decay can be accompanied by the emission of different types of radiation, the most common being alpha, beta and gamma radiation. These are identified by the Greek letters α, β and γ respectively. But what are alpha, beta and gamma radiation?

Alpha, beta and gamma radiation

The term *alpha particle* is more often used than *alpha-radiation*. It has been shown that alpha particles are identical with the nucleus of the helium atom. The helium nucleus contains two neutrons and two protons and these four particles are so tightly held together that the alpha particle can be treated as a particle in its own right. The alpha particle has a mass number 4 and carries two units of positive charge. We do not need to know what charge is, though we shall see later that it is a most important property of radiation.

There are two kinds of beta radioactivity—beta minus and beta plus. *Beta minus radiation* (β^-) consists of fast electrons. These electrons have identical properties to the electrons which are in orbit round the nucleus in atoms. It is not, however, one of these electrons which is ejected from its orbit in β^- decay, but an electron which originates in the nucleus before it is emitted. The particles carry one unit of negative charge.

Beta plus radiation (β^+) consists of particles which are analogous to beta particles except that they, like the alpha particles, are positively charged. The β^+ particles are often referred to as positrons. We can distinguish between β^+ and β^- particles by observing the way their direction is changed by a magnet. Positively charged particles are deflected in the opposite direction from negatively charged particles whereas uncharged particles are not deflected at all. In common usage the term beta radiation refers to β^- particles.

Gamma-radiation (γ-*rays*) belongs to the general class of radiation known as electromagnetic radiation. Other members of this class are light and radio waves, but gamma-radiation is much more penetrating than these radiations. Gamma-radiation is uncharged and is emitted as a result of changes within the nucleus.

Radiation sources

Radioactive material is usually identified by the principal radiation it emits. If it emits only alpha or beta particles, this is fine and we refer to alpha and beta sources. Very often, however, more than one type of radiation is emitted from the same piece of material. For example, both gamma- and beta-radiation is emitted from a piece of caesium containing the nuclide ^{137}Cs. In this case the source is sometimes referred to as a beta source and sometimes a gamma source, depending on the construction of the source. Charged particles are very easily stopped so we can construct the source in such a way that only the gamma rays escape from the source and present a radiation hazard. In this case, the source is referred to as a gamma source. If, however, the construction of the source permits both the beta particles and the gamma rays to escape then the source is referred to as a beta source. A similar situation arises where alpha particles and gamma rays are emitted from the same source. *Beta and alpha sources very often emit gamma-rays as well.*

In general, for sealed sources where the hazard is external to the body, gamma sources are more hazardous than beta sources which in turn are more hazardous than alpha sources.

Bremsstrahlung

There are additional complications when dealing with β^- or β^+ sources. When fast-moving electrons are slowed in materials they emit radiation called *bremsstrahlung* or braking radiation. There is always some *bremsstrahlung* emitted from a beta source, but there are also specially designed sources referred to as *bremsstrahlung sources* where this is the main type of radiation emitted from the source. X-rays, of which we have all heard, are just *bremsstrahlung*. In x-ray machines, electrons are accelerated until they are moving very fast. They are then stopped in a target and the x-rays are emitted. *Bremsstrahlung* or x-radiation, is identical in its properties to gamma-radiation. A γ-ray or x-ray of the same energy will behave in exactly the same way. *Hence* bremsstrahlung *sources should be regarded as gamma sources from the point of view of radiation protection.*

The complication with β^+ sources is that when the β^+ particle is stopped in material it combines with any electron it can find, is destroyed and radiation is emitted which is again similar in its properties to gamma-radiation. This is called annihilation radiation. All that need be remembered is that *it is best to regard β^+ sources as gamma sources.*

Electron capture and neutron sources

There is a process which occurs called *electron capture* (written as EC) in which an electron from an orbit close to the nucleus is absorbed into the nucleus. A rearrangement of the atomic electrons then results in the emission of characteristic x-rays. These x-rays have similar properties to *bremsstrahlung*; the different name is only used to identify the origin of the x-rays. Usually electron capture is followed by gamma-ray emission so it is best to regard an electron capture source as a gamma source.

We also encounter *neutron sources*. The material forming a neutron source emits mainly neutrons although from some neutron sources gamma-radiation is emitted as well. Since neutrons are biologically very damaging all neutron sources should be handled with very great care. There are three main types of neutron source, referred to as alpha-neutron, gamma-neutron and fission-neutron sources respectively. These sources find uses in the simultaneous measurement of the moisture content and density of soils and in medicine.

Alpha-neutron sources are made by mixing alpha-radioactive material with beryllium, although sometimes fluorine or boron is used in place of the beryllium. The alpha particles emitted by the radioactive material bombard the beryllium nuclides. A nuclear reaction takes place and neutrons are emitted when the beryllium nuclides change into nuclides of carbon. The neutron is uncharged. The process of nuclides being changed by bombardment with other particles is referred to as a nuclear reaction and this is what is implied when we talk of Cockroft and Walton splitting the atom.

Gamma-neutron sources are formed by mixing gamma-radioactive material with beryllium. The gamma rays induce the nuclear changes. These sources present a serious gamma hazard in addition to the neutron hazard.

We also encounter, though rarely, *fission-neutron sources*. There are several heavy nuclides, notably nuclides of uranium and thorium, which undergo spontaneous fission. In this process, the nucleus splits into two and when it does so neutrons are emitted. Californium-252 sources of this type are being increasingly used.

Neutron sources are usually measured in terms of the number of neutrons they emit. Thus an alpha-neutron source might have an output of 10 million neutrons/s. Since 14 neutrons/s, of high energy, falling on 1 cm^2 of the body exceeds the tolerance limit, the need for caution in using such sources is obvious.

It should be stated that there is no way of telling by appearance if a source is an alpha, beta or gamma source, although in the next chapter we shall see that we can make a shrewd assessment as to the nature of a source from the container in which it is normally kept. In figure 3.1(a), for example, alpha, beta and gamma sources are shown and it is not possible to tell by appearance which is which.

The energy of the radiation and energy units

One of the most important properties of any radiation is its energy. In the decay process each change of state of the nuclides is accompanied by a loss of mass and it is one of the firmly established principles of modern physics that energy and mass are equivalent. Most people meet energy units when they pay their electricity bill. Here the unit of energy used is the kilowatt hour (kWh), which is sufficient electrical energy to keep a one-bar fire burning for one hour. More often the unit of energy used is the joule since this has been adopted by the General Conference on Weights and Measures (1960). There are 3 600 000 J in 1 kWh so the joule is a very small quantity of energy. The energy associated with radiation is usually measured in yet another unit the mega electron volt, written as 1 MeV; 1 MeV is only equal to a tiny fraction of a joule and in fact would only keep a one-bar fire burning for approximately 0·0000000000000001 s.

Radiation has usually only a few MeV of energy associated with it. This is a small amount of energy, but nevertheless it is when the energy associated with radiation is deposited in the tissue of our bodies that damage occurs.

The symbol MeV or its sub-unit keV (kilo electron volt) should be remembered since they recur later in this book.

$$1 \text{ keV} = \frac{1}{1000} \text{ MeV}$$

Directly and indirectly ionizing particles

The transfer of energy from atomic radiation to matter, such as the tissue of the body, results in a phenomenon called ionization. It is this phenomenon which distinguishes atomic radiation from other forms of radiation such as heat and light. *Atomic radiation is therefore usually referred to as ionizing radiation.*

Alpha and beta particles (β^+ or β^-) are charged particles and in passing through a piece of material they collide with the electrons of the material making a series of successive collisions and losing part of their energy to each electron with which they collide. In this way the charged particles quickly lose all their energy and if the material is sufficiently thick they are brought to a halt, otherwise they all pass through the material but with a reduced energy. If the energy transferred to the electrons causes them to be knocked free from their atoms we say the atoms have been ionized and the process is referred to as ionization. *All charged particles can ionize in this way and are said to be directly ionizing.*

Gamma-rays and neutrons cannot ionize directly in this way. Gamma-rays collide with the electrons in materials, just like charged particles. However, in general each gamma-ray collides with only one electron, transferring all or part of its energy to that electron, or else it passes right through the material. If a sheet of lead is placed in the path of gamma-rays only some of the gamma-rays will pass through. This is in contrast to charged particles where either all the particles pass through or all are stopped. The ratio of the number of gamma-rays incident on a piece of material to the number passing through is referred to as the attenuation factor. For example, if 100 gamma-rays are incident on a piece of material and 20 pass through, the attenuation factor is 5($100/20 = 5$).

The different manner in which charged particles and gamma-rays interact with materials is in some ways analogous to a man running into a wood in the dark. If the wood was very dense, the man (charged particle) would simply bounce from one tree to another (the electrons) and soon be stopped. Or if the wood was not very thick he would emerge from the other side running more slowly with less energy. If, however, the trees were well spaced, the man, in this case analogous to the gamma-rays, could pass right through the wood unless he ran straight into a tree and was knocked down or at least had his direction changed as a result of the collision with the tree. The electrons in materials always appear to be well spaced to an incident gamma-ray but very closely packed to an incident-charged particle.

Neutrons collide with the nuclei of the atoms from which materials are formed and not with the electrons. They can induce nuclear reactions in which case the neutron is absorbed or captured into the nucleus and ceases to be a separate particle. Very often, however, protons or alpha particles are produced following the neutron capture. These charged particles are then stopped in the material ionizing the atoms of the material as discussed above. Hence the energy of the neutron is transferred to the material in which it is captured.

Neutrons can also transfer some of their energy to the material by 'bouncing off' the nuclei in a manner similar to two billiard balls. In this case the neutron is not absorbed but continues on its way with less energy. For example, the transfer of energy from neutrons to the tissue of the body is the result of the neutrons colliding with the protons which form the nuclei of the hydrogen atoms present in the tissue. Hydrogen and oxygen are the main elements present in tissue. The protons lose their energy in the tissue. Since neutrons and gamma-rays can release either charged particles or electrons in materials and since these can then ionize the materials, neutrons and gamma-rays are said to be indirectly ionizing.

Things to remember from chapter 4

(1) We encounter alpha (α), beta (β) and gamma (γ) sources.
(2) Beta plus (β^+) and electron capture sources (written as EC) are also met with. These should be regarded as gamma sources from the point of view of radiation protection.
(3) In general, more than one type of radiation is emitted from each source.

(4) Sealed gamma sources are generally more hazardous than sealed alpha or beta sources.

(5) Beta particles when slowed down emit *bremsstrahlung*. These x-rays have similar properties to gamma-rays and hence *bremsstrahlung* sources should be regarded as gamma sources.

(6) Neutron sources are particularly hazardous. Their output is given in terms of neutrons/s and not in curies.

(7) Associated with each radiation is an energy given in MeV or keV where $1 \text{ keV} = \frac{1}{1000} \text{ MeV}$.

(8) Atomic radiation is usually referred to as ionizing radiation. Alpha and beta particles are directly ionizing. Neutrons and gamma-rays are indirectly ionizing.

5

The shielding of radiation

Rules to be followed when working with radioactive materials

There are three simple rules which should be obeyed when working with radioactive materials.

(1) Keep as far away from the radioactive material as possible.

(2) Only stay near the radioactive material for as long as is absolutely necessary.

(3) Where possible shield oneself from the radiation.

These rules should be applied with all sources of radiation, whether the source is an x-ray generator or an open or sealed source. With open sources there is a fourth rule. *The open source material should be contained to prevent contamination.*

To obey the first two rules requires no special knowledge. Tongs or tweezers are used as necessary to keep the radioactive material as far from the hands and body as possible. This is illustrated in figure 3.2 which shows open source material being dispensed. In addition, the principle of containment is being applied. The operation is carried out over a large tray lined with absorbent paper so that if a spill occurs, the spillage will be confined to the area of the tray. It is to assist with this containment that open sources are usually handled in special laboratories as discussed in chapter 2. To obey the third rule, however, it is necessary to know what materials to use to shield each type of radiation mentioned in chapter 3. The following sections are therefore devoted to the subject of shielding.

Shielding

Table 5.1 summarizes the shielding materials which should be used with each type of source.

Table 5.1   The shielding materials to use with different types of source

Type of source	Shielding material
Neutron	Between 15 to 30 cm of water, wax or polythene, followed by 1 mm cadmium sheet or 1 cm of boron. When boron is used it is normal to mix boric oxide into the wax or water
Gamma, *bremsstrahlung*, x-ray	Lead; the thickness of lead depends on how active the source is and on the energy of the radiation emitted by the source
Electron capture and β^+ sources	These are best regarded as gamma sources
β^- sources	Usually 1 cm of Perspex is used as the shield followed by 1 mm of lead. The lead is used to shield the *bremsstrahlung* emitted as the beta particle is slowed down. For low-activity sources, where little *bremsstrahlung* is produced, the lead may not be required

Alpha particles are so easily stopped that they never present a shielding problem. A thin piece of paper is sufficient to stop most alpha particles. Because of their short range, alpha particles cannot penetrate the dead layer of skin surrounding the body although they would enter the eyes. It is the very short range of alpha particles which makes sealed alpha sources the least hazardous to handle. If an alpha source is more than 5 cm from the body the alpha particles will be stopped before they reach the body.

Alpha sources are shielded using very thin layers of material.

Beta particles are more penetrating than alpha particles. Beta particles will not be stopped by normal clothing. Nor will they be stopped by the dead layer of skin surrounding the body. They can penetrate up to 3 m of air. Hence shielding must be used with beta sources. As discussed in chapter 3, when a beta particle is slowed down it emits *bremsstrahlung* and since the production of *bremsstrahlung* is much reduced in materials of low mass number, it is customary to use Perspex or aluminium when shielding beta sources; 1 cm of material is sufficient. If necessary, this can also be clad with thin lead (1 mm) to shield the *bremsstrahlung*.

Beta sources are shielded using Perspex, clad in thin lead.

It is the fact that beta sources are fairly easily shielded and that at distances in air greater than 3 m the beta particles will not reach the individual, that makes sealed beta sources less hazardous to handle than gamma sources.

Gamma sources are shielded using lead.

The thickness of lead required depends on both the energy of the gamma-radiation emitted by the source and on the activity of the source. For highly active sources emitting high energy gamma-rays as much as 20 cm of lead may be required. It is the difficulty in shielding gamma sources which makes them the most hazardous to handle. We cannot rely on the air between us and the source or on our clothing to shield us. The shielding used with β^+ sources or electron capture sources is the same as for gamma sources.

Neutrons, like gamma-rays, again have a very long path length indeed in air and thus shielding is required. However, lead is not used as the shielding material. In this case the shielding is usually wax or polythene followed by a layer of boron of the order of 1 cm thickness,

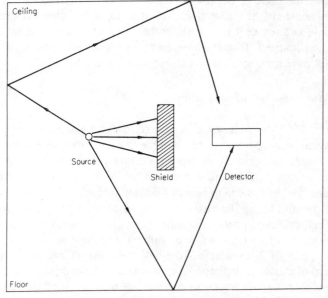

Figure 5.1   Illustrating the effects of scattered radiation

Table 5.2 Approximate half value thickness of lead for different gamma-ray sources

Gamma source	Half thickness lead (cm)	Gamma energy (MeV)
Cobalt-60	1·6	1·17, 1·33
Caesium-137	0·9	0·66
Iridium-192	0·25	0·30–0·61
Thulium-170	0·1	0·052, 0·084

or cadmium sheet 1 mm thick. Sometimes the boron and wax are mixed to form a borated wax mixture. Water itself forms an effective neutron shield especially if it has boron dissolved in it. The thickness of wax or polythene depends on the neutron energy but is usually about 15 to 30 cm thick.

Scattering

For complete safety, it is necessary when shielding a source to provide shielding on all sides to avoid the effects of scattering. Study figure 5.1; in this figure, the source is emitting radiation in all directions. The radiation travelling directly towards the detector is unable to reach the detector because of the shield. However, radiation leaving the source in other directions will strike the walls, floor and ceiling of the laboratory or it may strike some object in its path and have its direction changed. Thus if a complete shield is not used with a source, careful measurements for scattering must be made.

Half thickness or half value layer

In table 5.1 the thickness of lead used to shield gamma sources was not given. Instead it was stated that the thickness required to produce any desired shielding effect depends on how active the source is and on the energy of the gamma-rays emitted by the source. Let us now consider just how the thickness of shielding is chosen. This can best be done by introducing the quantity—*half thickness or half value layer*. The half thickness of any shielding material is the thickness which is required to reduce the number of gamma- or x-rays at some location by a factor of 2. In table 5.2 the half thickness of lead is given for several of the more common gamma sources. From this table it is apparent that the half thickness depends on the energy of the gamma-rays being emitted by the source and is greater for high-energy gamma-rays than low-energy gamma-rays. This is because high-energy gamma-rays are more penetrating than low-energy gamma-rays. The half thickness is also different for different materials. For example, the gamma-rays from a cobalt-60 source have a half thickness in lead of 1·6 cm but a half thickness in water of approximately 16 cm. Thus 1·6 cm of lead or 16 cm of water have the same shielding effect.

One half thickness reduces the intensity of gamma-rays by a factor of 2.

Two half thicknesses reduce the intensity of gamma-rays by a factor of 4.

Three half thicknesses reduce the intensity of gamma-rays by a factor of 8.

Thus each time a half thickness is added to a shield, the intensity of the gamma-rays at some point outside the shield is reduced by a factor of 2.

Example

Calculate the reduction in the intensity of the gamma-rays emitted by a caesium-137 source, if the source is placed in a lead container 2·7 cm thick, at some point outside the container.

The half thickness of lead is 0·9 cm for these gamma-rays. The number of half thicknesses in the shield is $3\left(\dfrac{2·7}{0·9} = 3\right)$ so the reduction in intensity is $(2 \times 2 \times 2)$ which is equal to 8.

It should be noted that this reduction in the number of gamma-rays arriving at some location outside the container might not be sufficient to make that a safe location. With a highly active source emitting very many gamma-rays, more shielding might be required to reduce the intensity of the gamma-radiation to an acceptable level.

In fact the picture is more complicated than is presented here. It is also necessary in using half thicknesses to decide if the gamma-rays will be totally absorbed in the material forming the half thickness or simply have their direction changed. In practice 4 cm of lead is usually sufficient to shield most gamma sources.

We can use the idea of half thickness with neutrons. For example, for 8 MeV neutrons, the half thickness of water is approximately 7 cm. Hence 14 cm of water between some object and a source emitting 8 MeV neutrons will reduce the number of neutrons falling on the object to one-quarter of the value when the water is not there.

The idea of half thickness is not used with charged particles since, as was mentioned in chapter 4, charged particles are either stopped in material or else they all pass through. Thus if 10 alpha particles fall on a piece of paper, all the alpha particles will pass through but with a reduced energy which means they will be moving more slowly. When the paper becomes a certain thickness all the alpha particles will be stopped in the paper—the thickness is called the range of the alpha particle for that material. The situation is similar for beta particles but the range is not so clearly defined.

Things to remember from chapter 5

(1) Alpha particles are stopped by very thin layers of material so the air surrounding an alpha source provides good shielding.

(2) Beta sources are shielded using Perspex followed by lead.

(3) Gamma-ray sources, β^+ sources and electron capture sources are shielded using lead.

(4) Neutrons are shielded using light materials like wax or water to slow the neutrons down; the neutrons are then absorbed using boron or cadmium.

6

Radioactive decay

The half-life

If you were given a piece of radioactive material, you would probably wonder if the activity of the material would always remain the same or if the activity would decrease from day to day. In fact,

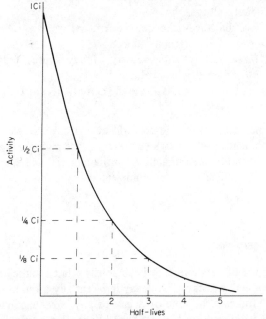

Figure 6.1   Illustrating the decrease in the activity of a source with the half-life of the source. After one half-life, the initial activity of 1 curie has decreased to $\frac{1}{2}$ Ci. After two half-lives the activity is $\frac{1}{4}$ Ci and so on

early studies of radioactive materials showed that the activity of these materials decreases at different rates depending on the radioactive nuclides forming the material. For some materials the activity decreases very rapidly but for others the activity decreases very slowly. If, however, the activity of a piece of material decreases by some fraction over, say, a day, then it decreases by the same fraction each succeeding day. It was convenient to introduce a quantity called the *half-life* or *half-period* to describe this behaviour. The half-life was defined to be the time required for half the radioactive nuclides in any piece of radioactive material to decay. Half-lives were found to vary from fractions of a second to millions of years depending on the particular radioactive nuclide studied. The half-life of any particular radioactive nuclide, characterized by its mass number, was however always the same. A measurement of the half-life is very useful in identifying the active nuclides in any piece of radioactive material.

The half-life to associate with any particular source is important as is illustrated in figure 6.1, since

(a) in *one* half-life the activity of any source falls to *half* of its original value;

(b) in *two* half-lives the activity of any source falls to *one-quarter* of its original value;

(c) in *three* half-lives the activity of any source falls to *one-eighth* of its original value;

(d) for each additional half-life the activity decreases by another factor of *two*.

In fact, no radioactive material will ever decay completely; some radioactive nuclides will always be present in the material even though their number and hence the activity of the material will eventually be negligibly small. The time for complete decay cannot therefore be stated.

The decay constant

The half-life of a radionuclide is obviously related to the laws which control radioactive decay. The decay is essentially a random process. We cannot forecast the time of decay of any particular nuclide in a source. All we can do is to state the chance of a particular nuclide decaying in any given second. Since a radionuclide has no memory this chance remains the same until the nuclide eventually decays. The chance of decay is called the *decay constant*, λ, of the nuclide and it is the decay constant which determines if a source has a long half-life or not. If the nuclide has a very large decay constant, so that in a source very many of the nuclides are decaying per second, then clearly the source will have a short half-life since all the nuclides will quickly become inactive. If, on the other hand, the nuclide has a very small decay constant then very few nuclides will decay per second, so that the source will remain active for a very long time and have a long half-life. If the half-life associated with any radioactive material is short, and hence the decay constant is large, then there will be a large number of nuclides in the material decaying in any one second. Now the activity of any piece of radioactive material is a measure of the number of disintegrations/second (d/s) and it therefore requires a very small quantity of material indeed to have an activity of 1 curie (1 Ci)—1 Ci, you will remember, was defined to be a disintegration rate of 37 000 000 000 d/s. Thus as little as one-millionth of a gram, a tiny speck of material, can have an activity of 1 Ci. *Serious contamination can result from spillage of exceedingly small quantities of materials indeed.*

Knowing the half-life to associate with any radioactive material is very important. Suppose, for example, some liquid containing caesium-137 is spilled. Since caesium-137 has a half-life of 30 years, it would take 30 years for the activity of the spilled material to decrease to half of its original value. So we would have to take steps to remove the contamination. On the other hand, if some manganese sulphate, containing manganese-56, was spilled it might be better simply to isolate the area and wait for the activity of the material to decay. Manganese-56 has a half-life of approximately $2 \cdot 5$ h, so in 25 h the activity will have decayed to just over one-thousandth of its original value. In another 25 h, the activity will be decreased to $(\frac{1}{1000} \times \frac{1}{1000})$, one-millionth of its original level and so on.

In dealing with contamination, if the material has a short half-life it is often best simply to isolate the area where the contamination occurs and wait for the activity of the material to decrease.

It is because the half-lives of nuclides like uranium-238 and thorium-232 are so long, even when compared to the age of the earth, that the soils or rocks containing them are still radioactive today, millions of years after their formation. The nuclide potassium-40, found in the bones of our bodies, has a half-life of 1 250 000 000 million years so the bones in our bodies will be radioactive long after we are dead. The age of meteorites, rocks, archaeological artefacts and biological materials can be obtained by measuring the activity of the radioative nuclides they are known to contain.

Waste disposal

A new problem has to be faced as a result of the increasing use of nuclear power stations—the disposal of radioactive wastes. The materials forming the waste produced from nuclear power stations are in gaseous, liquid or solid form and if the half-lives associated with these waste materials were of the order of hours or even a few days their disposal would present no difficulty. They could simply be stored until the activity reached a safe level when they could be disposed of just like any other waste. Unfortunately, some of the waste products are made up of nuclides which have half-lives as long as 30 years, and since these wastes are also the most active, it will be many centuries before they can be disposed of safely. Storage is therefore only a temporary solution. This type of waste arises mainly as a result of reprocessing the uranium used as the fuel in reactors. In the reactor the uranium nuclides split into two, forming lighter nuclides which are referred to as fission products. After a time, the fuel has to be removed and reprocessed. Valuable unused uranium-235 and also plutonium, which is created in the fuel, are recovered in the reprocessing, but highly active solutions containing the fission products are produced and it is these which present the greatest difficulty. What should be done with them? At present the practice is to reduce their bulk by evaporation and then they are transferred to special tanks in large concrete vaults. Even though the tanks are double-walled, there have been a few cases where the material has leaked into the surrounding ground. This is somewhat disturbing when it is remembered that the tanks must last for centuries. This difficulty with the time for which the wastes must be stored, rather than the quantity of the waste itself, is one of the causes of controversy over nuclear power. In fact, the volume of the waste or its weight is fairly small when compared with the volume of waste from coal-fired power stations.

Waste material arises not just from nuclear power stations. On a much smaller scale it arises in hospitals, industry, research and teaching institutions. Here the solid waste is usually disposed of along with ordinary trade waste. There are, though, strict limits placed on how much waste may be disposed of at any given time, taking into consideration the activity of the waste. Liquid waste can be disposed of as sewerage, but the activity is again carefully controlled.

Things to remember from chapter 6

(1) Every radioactive nuclide has a half-life associated with it. This half-life can be used to estimate how long the activity of a particular source will take to fall to a safe level.

(2) In *one* half-life the activity of any source will fall by a factor of *two*, in *two* half-lives by a factor of *four* and so on.

(3) In dealing with contamination, if the material has a short half-life, it is often best simply to isolate the area where the contamination occurs, and wait for the activity of the material to decay.

(4) *Serious* contamination can result from spillage of *exceedingly small* quantities of material.

7

Radiation units

It is clearly important to have some understanding of the units which are used in radiation protection. The instruments which are discussed in chapter 8 all have a dial or scale which is graduated in terms of one or other of these units, and in order to make proper use of these instruments it is necessary to understand what any particular reading on the scale means. One of the most frequently used units is the *rem*. This was used in chapter 1 when the concept of maximum permissible dose was discussed, and was also used when the annual dose to each individual from the natural background was given. The other two units which are most frequently used are the *roentgen* and the *rad*. Use is also made of rad/h, rem/h or roentgen/h to describe the rate at which radiation is received.

Let us consider each of these units in turn.

The roentgen

The roentgen (abbreviated to R, or sometimes r) was the first widely used radiation unit. It is used with gamma- or x-radiation only. The roentgen was originally defined to be the quantity of gamma- or x-radiation which produces a specified quantity of ionization in air. The quantity of ionization produced by the radiation can be measured in a special instrument called an *ionization chamber*. The modern usage of the roentgen, however, is as the unit of exposure where exposure is the quantity used in expressing the amount of ionization produced in air by gamma- or x-radiation. Thus an exposure of 10 R implies more gamma-rays or x-radiation than an exposure of 1 R. In fact, the most important thing to remember about the roentgen is that since it is used only with gamma- or x-radiation, *any instrument with a scale marked in roentgens will be intended primarily for use in measuring gamma- or x-radiation.*

The rad

When radiation is absorbed in tissue some fraction or all of the energy associated with the radiation is transferred to the tissue. Radiation damage may result because of this transfer of energy and it is found that the greater the transfer of energy, the more likelihood there is of damage. The energy absorbed per g or kg is called the absorbed dose and the absorbed dose is expressed in a unit called the rad (radiation absorbed dose) which can be used with all types of radiation. One rad is defined as an energy deposition of 62 500 000 mega electron volts per gram which is written as $6 \cdot 25 \times 10^7$ MeV/g, and thus 1 mrad $= \frac{1}{1000}$ rad, is equivalent to 62 500 MeV/g.

It is best to express the rad in terms of MeV/g since the energy associated with nuclear radiation is always given in MeV.

A new unit called the *gray* (abbreviated Gy) has recently been adopted as the unit of absorbed dose and this will gradually replace the rad.

1 gray is equivalent to 100 rads

Harmful levels of radiation are often stated in rads. For example, over 100 rads must be imparted over a short period to a substantial part of the body before most individuals will show significant clinical symptoms.

The rad is clearly a more useful unit than the roentgen since it can be used with all types of radiation. The rad expresses the energy absorbed in 1 g of any medium which is the critical factor in determining the risk involved in any radiation exposure.

We can understand the rule of thumb given in chapter 2 that curie sources are more hazardous than mCi or μCi sources in terms of the absorbed dose, since in any given situation the absorbed dose from a curie source will be 1000 times greater than the absorbed dose from a mCi source and so on. Again, when we are considering sources external to our bodies, alpha sources will clearly be the least hazardous because of the ease with which they can be shielded and the absorbed dose reduced to zero. It is because we can do very little to reduce the absorbed dose once radioactive material is inside our bodies that we handle open source material with extra care. Clearly the shorter the time we remain in the vicinity of a source the smaller will be the absorbed dose. Hence the rule

Remain in the vicinity of a source for as short a time as possible.

All measures taken in radiation protection are designed to reduce the absorbed dose in the tissue of our bodies.

The rem—dose equivalent

Although the rad is a most useful unit, it is found that in biological systems the degree of damage produced by a given dose depends on many factors including the type of radiation and how the radiation is distributed. To take this difference into account, when adding the absorbed dose of different radiation we must multiply the absorbed dose of each type of radiation by a quality factor (QF) which reflects the ability of the particular type of radiation to cause damage. The quantity obtained when the absorbed dose in rads is multiplied by a quality factor is known as the dose equivalent or rem. Following the International Committee on Radiological Protection this is most often called the dose, as in chapter 1 for example.

Thus dose equivalent (rem) is obtained by multiplying the absorbed dose in rads by the appropriate quality factor. The quality factors to use are given in table 7.1. These quality factors are recommended by the I.C.R.P.

Table 7.1 Use of quality factors

Type of radiation	Quality factor	Energy of radiation
X-rays, γ-rays, β particles	1	All energies. For very low energy β particles with energy not greater than 0·03 MeV the QF is increased to 1·7
Slow neutrons	2·5	Energy less than 0·01 MeV
Fast neutrons, alpha particles	10	All energies of alpha particles

Let us see how the rem is used in any given situation.

Suppose you return a film badge, discussed in the next chapter, (page 44), for measurement. In the United Kingdom you will receive a form specifying the gamma dose and beta dose in rem. These will simply be the absorbed dose in rads multiplied by the quality factor.

For beta- and gamma-radiation, the quality factor is 1 and therefore for beta- and gamma-radiation, rem and rad are synonymous. For gamma-radiation an exposure of 1 R is equivalent to an energy transfer of very nearly 1 rad in tissue, *so that for gamma-radiation the units rem, rad and roentgen can be regarded as equivalent.*

For alpha particles or neutrons the rem and rad are not synonymous. Suppose a radiation worker received an absorbed dose of 1 rad from alpha radiation and an absorbed dose of 10 rads from neutrons. What is his dose equivalent?

Dose equivalent = absorbed dose × quality factor

The dose equivalent for alpha-radiation is 1 × 10 and for neutron-radiation is 10 × 10, since the quality factor for neutrons and alpha particles is 10. The worker's total dose or dose equivalent is thus 110 rem, which would be far in excess of his permitted annual dose of 5 rem. The International Committee on Radiological Protection takes a lot of trouble to point out that though the imposition of limits to the dose equivalent (or dose which can be received by any individual per year) limits the risk involved in being exposed to radiation, the dose is not a direct measure of the risk or even of the effect of the limitation. The risks are not the same when different organs receive the same dose.

Alpha particles have a large quality factor associated with them, so an absorbed dose of alpha-radiation is likely to produce a greater biological effect than an equal absorbed dose of beta radiation. This is reflected in the radiotoxicity of nuclides which emit alpha particles. Most nuclides which emit alpha particles are classified as being very highly toxic.

Hence when handling open source materials, those which emit alpha particles must be handled with the greatest care.

With internal sources, in certain cases, the dose equivalent in rem has to be further modified by a factor which takes into account any non-uniform distribution of the absorbed dose within the body. For external sources, the factor is 1 and can be ignored.

Dose rate

The roentgen, the rad and the rem are the units of exposure, absorbed energy/g and dose equivalent respectively. However an absorbed dose of 10 rad or a dose equivalent of 10 rem may be received over different times and in specifying radiation areas, it is the rate at which radiation will be received which is important. The dose rate is expressed in rem/h, the absorbed dose rate in rad/h and the exposure in R/h.

For example, radiation areas must be clearly identified if the dose rate in the area exceeds 0·75 mrem/h, where 1 mrem as usual equals $\frac{1}{1000}$ of a rem.

Flux

The measurement of the absorbed dose from neutrons is such a difficult measurement to make that the I.C.R.P. has recommended that a quantity called the *flux* be used with neutrons and that the neutron flux be related to the dose equivalent (rem) as illustrated in table 7.2. The neutron flux is simply the number of neutrons falling on each cm² of the body/s. From table 7.2 we can see that 12 neutrons of energy 14 MeV falling on each cm² of the body results in a dose rate of 0·75 mrem/h.

Table 7.2   Neutron flux which corresponds to 0·75 mrem/h

Neutron energy	Flux (neutrons $cm^{-2}\ s^{-1}$)
Thermal (slow)	680
1 MeV	19
10 MeV	17
14 MeV	12

The flux is reduced by a factor of 4 each time we double the distance between ourselves and a source. The smaller the flux, the fewer the number of particles which enter our bodies and deposit energy, so it is clearly best to stay as far away from a source as possible. This was the first rule given in chapter 5.

There is obviously a relationship between the activity of a source given in curies, or in the case of a neutron source, the source strength in neutrons/s, and the dose rate at any specific distance from the source. The activity determines the number of particles arriving at any distance from the source, each with a precise energy associated with them. The greater the activity, the greater will be the dose rate at any point. This was the basis of the rule of thumb given at the end of chapter 3.

Things to remember from chapter 7

(1) In radiation protection the units most commonly used are the rem, the rad and the roentgen.

(2) The rad is the unit of absorbed dose. An absorbed dose of 1 rad corresponds to an energy deposition of 62 500 000 MeV/g of material.

(3) The rem is the unit of dose equivalent, 1 rem = 1 rad × the quality factor.

(4) The quality factor is different for different types of radiation and allows for the fact that equal absorbed doses in rads from different kinds of radiation do not produce the same degree of biological damage.

(5) The roentgen is only used with gamma- or x-radiation and is the unit of exposure—the quantity used to measure ionization produced by gamma- or x-radiation in air.

(6) For gamma-radiation 1 rad = 1 rem = 1 roentgen.

8

The detection and measurement of radiation

In the day-to-day control of radiation there are, in general, three types of measurement which must be made. Measurements are required of the dose accumulated by each individual in the course of his or her work, to ensure that the maximum permissible dose is not being exceeded. The dose rate or absorbed dose rate in areas has to be measured so that radiation areas, where the dose rate exceeds 0·75 mrem/h, can be identified. It is also necessary to detect and measure any contamination of surfaces and clothing, or even contamination of the air within a laboratory. This ensures the adequate control of open source material. Instruments have been developed with each of these tasks in mind and these instruments are considered in turn in the following sections of this chapter.

Personnel dosimetry systems

These instruments measure the dose accumulated by individuals over any given period of time. The radiations of greatest external hazard are gamma- and x-radiation, β-radiation and neutrons so the dosimetry systems in use are designed to measure the dose from these radiations. The systems in current use are described below.

The film badge

This is the most common dosimeter used in monitoring whole body dose. Photographic emulsions which have been exposed to radiation appear blackened after development. The degree of blackening is dependent on the absorbed dose received by the film and hence the dose can be read off from a dose-calibration curve. This dose in rads can then be converted to rems by using the appropriate quality factor. The film is worn in a specially designed holder, which contains a number of strips of plastic and metal, called filters, so that beta-, gamma- and x-radiation doses can be measured. The dose from very low energy neutrons, called thermal neutrons, can also be measured. In the United Kingdom a film badge service is provided by the Radiological Protection Service. Dose equivalents are measured for β-rays, of energy > 0.25 MeV, for gamma- and x-rays between 20 keV and 2 MeV, and for slow neutrons or for mixtures of these radiations. The accuracy achieved is about ± 20 per cent. The badge is worn in its holder at some convenient place on the body. *For fast neutrons, it is necessary to wear a second film badge.* This is because fast neutrons produce no blackening of photographic film. Instead we infer the number of neutrons entering a film badge by measuring the number of tracks produced in emulsions which are placed on each side of a layer of material containing hydrogen.

Thermoluminescent dosimeters

Certain materials after being exposed to radiation emit light when heated. This fact is made use of in thermoluminescent dosimeters. The dose received is derived from a measure of the quantity of light emitted, when the material is heated. They are less affected by environmental conditions than film badges but unlike film badges they cannot readily distinguish between mixed radiations. Hence both systems are used to complement one another. For example, finger-stalls of thermoluminescent materials are often used if that is the area of the body particularly at risk, as it is for example in x-ray crystallography work.

Pocket dosimeters

Neither of the above systems is direct reading. The dose is only obtained after processing of the film badge or thermoluminescent material and this takes time. Therefore under conditions where the likely dose is high and where it is necessary to have a continuous visual indication of the dose, a system is used where the dose can be read directly as it is received by pocket dosimeters, and these are used

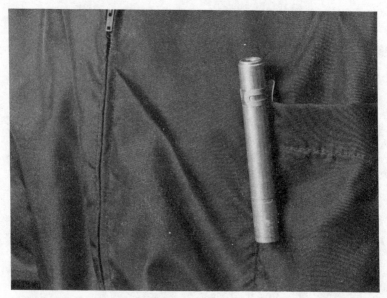

Figure 8.1   A typical pocket dosimeter

in addition to film badges where for example a planned emergency exposure of a radiation worker is undertaken. A typical pocket dosimeter is shown in figure 8.1. A dose can be read directly as it is accumulated by viewing the deflection produced in a quartz fibre. The deflection is produced as a result of ionization of the gas in the dosimeter by gamma- or x-radiation. The scale is marked in roentgen since the quantity being measured is the ionization in air by gamma- and x-radiation. The dosimeters are available with sensitivities ranging from 200 mR to thousands of roentgen.

Clearly as indicators of whole body dose, all the personnel monitoring systems are to some extent limited since beams of radiation which do not pass through the monitors would not be recorded.

Instruments for measuring dose rate and contamination

It is not possible to describe in detail all such instruments. All we can do is outline their basic features and the basic methods used in measuring the radiation. The two most important things to remember are firstly *to choose the correct instrument for each type of radiation* and secondly *to make sure that the instrument is functioning properly*.

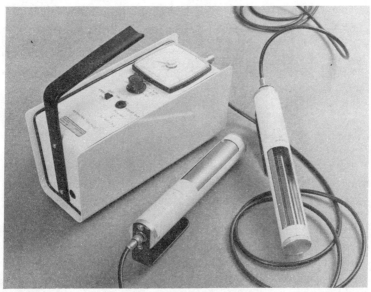

Figure 8.2   A β and γ geiger probe. With the metal blind closed only gamma-radiation is detected. This feature can be used to distinguish between gamma- and beta-radiation

The second point may seem obvious but in the author's experience very often instruments which are not working properly are used and relied upon. Most of the instruments are portable and rely on batteries to provide their power. They will not function if the batteries are flat, so a means is usually provided for testing the batteries. Again, if the reading on the scale of the instrument is to be relied on the instrument should be calibrated regularly. In the United Kingdom, a calibration service is provided by the National Radiological Protection Board.

The choice of which instrument to use with which type of radiation is usually related to whether the radiation can enter the detecting system or not. In figure 8.2 a typical instrument is shown. It is in fact a geiger counter which is discussed in more detail below. Before any radiation is detected it must pass through the glass wall of the counter so that it can collide with the gas filling of the counter. This particular counter therefore can detect gamma-rays which easily pass through the wall, and $β$-rays of energy greater than 0.25 MeV provided the metal blind is not in position. If it is in position, the $β$ particles stop in the blind and are not detected. Note that this enables us to distinguish between $β$- and gamma-radiation simply by taking measurements with the blind open and closed. This instrument cannot detect alpha particles since they cannot penetrate the glass wall of the counter. Thus it is essential to choose an instrument which lets the radiation into the detection system.

Let us now consider contamination monitors and dose rate monitors separately.

Contamination monitors

These are usually only used to detect the presence of radiation and hence the scale used in the instruments is normally in counts/s which gives a direct measure of the flux falling on the instrument. Calibration of the instrument is achieved by measuring the count rate/s from sources of known curie strength. A calibration is made for each type of radiation for which the instrument is used and for different energies of radiation since, for example, equal fluxes of 1 and 2 MeV gamma-rays will not produce the same counts/s from the instrument. Usually a calibration curve is provided by the manufacturer of the instrument, but periodical checks should be made to see that the calibration has not altered.

The detection system in these instruments is usually a geiger counter, a scintillation counter or a solid state detector.

The geiger counter consists of a conducting chamber which is filled with gas and which has mounted at its centre, but insulated from the chamber, a fine wire to which a high voltage can be applied. Charged particles ionize the gas and the resulting electrical current in the chamber can be measured. Very often the current is turned into a voltage pulse using electronic circuits and hence the number of pulses/s is a measure of the number of particles or gamma-rays entering the chamber. The instrument shown in figure 8.2 is a typical geiger counter which can be used for detecting β particles or gamma-radiation.

In the scintillation counter, the radiation falls on certain materials known as scintillators. The radiation produces a flash of light from the scintillator. The number of light flashes/s gives the number of incident particles or gamma-rays. The flashes are counted using a photomultiplier and electronic counting system. The purpose of the photomultiplier is to turn the light flash into an electronic pulse which can be measured. The instrument shown in figure 8.3 is a typical contamination monitor which uses scintillation counting. The scintillation

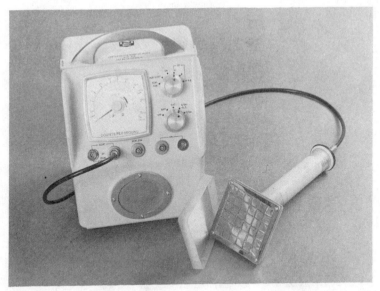

Figure 8.3 A typical contamination monitor. Depending on the measurements to be made different probes are used. The probe shown is a scintillation counter. To detect alpha-radiation the thin foil window must be exposed as shown

counter has the advantage over the geiger counter in that it is much more efficient for detecting gamma-rays. The efficiency of any counter is the number of pulses produced in the counter divided by the number of particles incident on the counter. In the geiger counter only about two per cent or less of all the gamma-rays entering the counter are absorbed and produce pulses. In the scintillation counter, the efficiency can be as high as 70 per cent for gamma-rays. In addition, the amount of light produced in the scintillating material is dependent on the energy of the radiation. Hence we can use the scintillation counter to identify nuclides by the energy of their radiation. For gamma-rays this is called gamma-ray spectrometry and is the usual method employed in identifying an unknown radioactive nuclide. The geiger counter produces a pulse whose size is independent of particle type and energy.

The solid state radiation monitors are normally made from hyper-pure crystals of silicon containing trace quantities of phosphorus, boron, gallium or aluminium. The radiation is detected by the ionization it produces. The advantage of this system is that it can be made

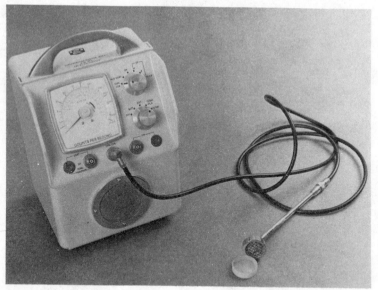

Figure 8.4   A solid state probe used for detecting alpha-radiation. The protective plastic cover is removed while the probe is in use

very small. Figure 8.4 shows a typical solid state probe which is used for measuring alpha-radiation.

The use of contamination monitors is supplemented by smear counting as an indirect monitor. A filter paper is wiped over a known surface area and the activity picked up on the filter paper counted. The activity measured can be related to the contamination level of the surface. Contamination monitors and smear surveys all establish the presence of contamination on surfaces.

For airborne contamination, the air is drawn through filter papers and the activity of the material absorbed in the filter paper determined. This measures the contamination which is in the form of very fine particles. If the material causing the contamination of the air is itself a gas, then the air is again filtered but the activity of the gas passing through the filter is measured.

Dose rate meters

These are very often called survey meters since they are used to measure radiation levels. Unlike contamination monitors these give a measurement in rad/h or rem/h. Since the absorbed dose is a measure

Figure 8.5   A β and γ dose rate meter. The scale is calibrated in mR/h. A true dose rate reading is only given for gamma-radiation. With the plastic cover removed the instrument can be used to detect beta-radiation but not to give a true dose rate for the β-radiation

of the energy deposited per g in the material, the system used must clearly measure this energy deposition. In general, the dose rate meters all incorporate a type of gas counter called an ionization chamber. Like a geiger counter this consists of a conducting cavity filled with gas and arranged so that when ionization occurs in the gas the resulting current can be measured.

In this instrument, however, the current is directly related to the energy deposited in the chamber so that if the chamber is lined with tissue-equivalent material, a direct measure of the absorbed dose in rads is obtained. This is easily converted to rem dose as discussed in chapter 7. Very often the chamber is not lined with tissue-equivalent material in which case the current is proportional to the ionization produced in the gas and hence the instrument can be used to measure the exposure of gamma- or x-radiation in roentgen. A typical instrument is shown in figure 8.5. This instrument is an ionization chamber which measures gamma- or x-radiation exposure rate in mR/h which can be taken as mrem/h. This instrument detects β-radiation if the end plastic cap is removed but it does not give the dose rate for β-

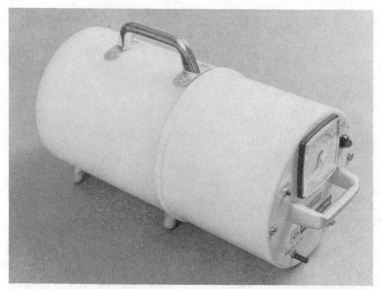

Figure 8.6   A neutron rem counter. The neutrons are first slowed down in the polythene before being detected

radiation. Geiger counters and scintillation counters are also used as dose rate meters although strictly speaking they do not measure the absorbed dose directly.

Neutron monitors

Neutrons require special detection methods as a result of the different way they interact with materials. The instrumentation used to measure gamma- or x-radiation will not detect the presence of neutrons.

If we wish to detect fast neutrons, we usually allow them to collide with atoms in light material to slow them down, after which their presence is detected by letting them collide with boron. Alpha particles are produced and the number of alpha particles counted. The number of pulses from the counter gives us a measure of the number of neutrons/cm^2 incident on the detector and we can calibrate the instrument using neutron sources of known intensity and energy. Using the relationship between neutron flux and dose equivalent as given in chapter 7, it is easy to convert the measured flux to the appropriate dose rate. Hence the scales are marked in rem/h.

For slow neutrons, gas counters are used directly with fillings of hydrogen or boron trifluoride. In colliding with hydrogen, the neutrons produce protons and in being absorbed by the boron they release alpha particles. Counting the number of protons or alpha particles released per second gives the neutron flux. Figure 8.6 shows a typical neutron counter which is calibrated in rem/h.

Measurement of internal radiation

With alpha and beta emitters the presence of radiation can only be detected by examining nasal swabs, the saliva and sweat, samples of blood among other things to detect the presence of radiation. For gamma-radiation which can escape from the body, whole body scintillation counting can be used. The subject is placed in a low background area and the gamma-rays detected using a scintillation counter.

Things to remember from chapter 8

(1) There are three main classes of instrument in radiation protection.
(2) Contamination monitors measure the presence of radiation. The scale on these instruments is usually in counts/s.
(3) Personnel dosimeters measure the total accumulated dose received by individuals over any given time, or in some cases the dose can be read off as it is being accumulated.
(4) Dose rate meters indicate in rad/h or rem/h the radiation level in any area and hence can be used to indicate how safe any area is.
(5) It is most important to choose the correct instrument according to the type of measurement to be made and the radiation to be measured or detected. Again it is most important to test that any instrument is functioning properly.
(6) The instruments used for neutrons are quite different from those used with other types of radiation.

9

Legislation relating to the use of radioactive materials

In most countries persons possessing, using, selling or transporting radioactive materials must be licensed by an appropriate authority. In the United States full information on federal and state licensing is available from

> U.S. Atomic Energy Commission,
> Division of Material Licensing,
> Washington D.C. 20545.

and in Canada from

> The Atomic Energy Control Board,
> Post Office Box 1046,
> Ottawa, Canada.

In the United Kingdom, registration must be made in Scotland with the Scottish Development Department in Edinburgh and in Ireland, England and Wales, with the Ministry of Housing and Local Government.

In this chapter, the laws and regulations which are in force in the United Kingdom are considered briefly along with the steps which must be taken to ensure they are adhered to.

Radioactive Substances Act (1960)

This Act was introduced to control the disposal and accumulation of radioactive waste. Under this Act, registration is required for open

54

and sealed sources, and the disposal of radioactive waste is prohibited without authorization. Certificates of registration are issued to each user showing the sealed sources, total quantity of open source material and amount of waste which can be held and disposed of respectively. The manner in which the waste must be disposed of is also stated. These certificates should be on display and place certain obligations on the user regarding the storage and handling of active materials. For example, when not in use the sources must be kept in a fireproof store. The general rules for handling radioactive materials as outlined in this book must be adhered to.

The ionizing radiation regulations and the various codes of practice

There are two sets of regulations under the Factories Act—the *Sealed Sources Regulations* (*1969*) and the *Unsealed Radioactive Substances Regulations* (*1968*). These regulations and codes of practice are intended to ensure the safety of all personnel working with radioactive materials. They have much in common, the chief difference being the requirement under the regulations to keep the local district factory inspector informed of the use of sources and of any accidents involving their use. Certain responsibilities are placed on the controlling authorities in the various organizations using radioactive materials or equipment which produces radiation.

Classified and designated persons

Firstly the authorities must appoint competent persons who will have special responsibilities for implementing the regulations or codes of practice. Secondly it is necessary to classify workers into two categories. Designated persons, who are likely to be exposed to radiation doses in excess of 3/10 of the maximum permissible dose of 5 rems, and non-radiation workers. All persons working with open source materials are defined as classified and designated persons. Designated persons will be subject to special supervision which will include

(a) Personnel monitoring as described in chapter 7.
(b) The recording of the radiation doses they receive.
(c) The provision of a transfer certificate on change of employment—this will show the total dose equivalent

received to date by the worker. Where appropriate a copy
should be sent to H.M. District Inspector of Factories.
(d) A pre-employment medical examination by a competent
medical adviser. This will be a general medical examination
but will include a blood count. The local Medical Officer of
Health will provide the name of a suitable doctor. A re-
examination is required annually if greater than 3/10 of the
permitted maximum dose per year is received, otherwise it is
not. The particulars of the examination should be entered by
the doctor in a Health Register.

The purpose of these requirements is to ensure that the maximum
permissible levels given in chapter 1 are not exceeded. Should a
designated worker receive a dose of radiation greater than the permitted level he should be suspended from work. Provided he has not
exceeded his permitted life dose as calculated using the formula given
in chapter 1, he may restart work at the beginning of the next calendar
quarter. If, however, the dose exceeds 25 rems, then reference should
be made to the competent technical authorities, as regards future exposure of the individual. It is probably best to consult the authorities
in all cases, as a precaution, and where appropriate the factory inspector must be notified. Records of the dose received by each worker and
the transfer certificates should be retained for 30 years.

*Designated persons should be given appropriate training concern-
ing the hazards involved and precautions to be observed when using
radioactive materials or radiation.* It is the responsibility of the
designated workers to give an employer his transfer certificate, if any,
when starting employment and he must submit to medical ex-
aminations as necessary.

*It is essential that at all times he wears the film badge or personnel
dosimeter provided and returns these as instructed.* He should also
notify the competent person of any suspected overexposure.

A designated person may be required to submit to biological
monitoring where necessary.

Clearly before it is possible to decide who should or should not be a
designated worker, it is necessary to assess the dose rate in any area
containing radioactive material. The regulations define a radiation
area as part of a factory in which any person is exposed to a radiation
dose rate which when averaged over any 1 min exceeds 0·75 mrem/h,
and requires that all persons employed in a radiation area be
designated as a radiation worker. The authorities must provide a dose

RADIATION
WARNING SIGNS

32.(1) There shall where reasonably practicable be a barrier or barriers marking the boundaries of every radiation area or where the use of such barrier or barriers is not reasonably practicable the said boundaries shall be marked by other suitable means.

(2) Suitable notices warning persons in the vicinity shall be displayed at a sufficient number of suitable places on or near to the boundaries of all radiation areas. (1969 No. 808 FACTORIES, The Ionising Radiation (Sealed Sources) Regulations 1969).

Size 13.3/4" x 9.3/4"

:::: Represents yellow

Figure 9.1   A radiation area warning notice

rate meter of the type described in chapter 8 which should be calibrated by a qualified person every 14 months but which should at all times be kept in proper working condition. Radiation areas should be clearly marked to prevent non-designated persons entering them. A typical warning notice is shown in figure 9.1. A radiation area should wherever possible be bounded by barriers and have limited access.

Precautions with sealed sources

Suitable arrangements must be made for the storage and accounting of these sources. In addition, they should be tested every 26 months for leakage which might lead to contamination, and records kept for a period of three years after the source has ceased to be used. These tests are performed in the United Kingdom by the National Radiological Protection Board. Records must be maintained of all sources held, and when received, and notification must be made in writing to the appropriate authority of the loss of any sealed source or of the breakage or leakage of a source.

At no time must a sealed source be manufactured from open source material. Radiation can affect the bonding properties of glues and other materials so that such sources are likely to leak eventually and spread contamination. Accidents have already been caused in this way.

Precautions with open sources

Open sources are, as was indicated in chapter 3, classified according to their radiotoxicity, and before they are handled they should be classified and the type of laboratory suitable for handling them prepared as given in table 3.1 (page 17). All the precautions are designed to prevent the intake of radioactive material into the body and to prevent contamination of clothing, surfaces, among other things.

Thus no unnecessary material should be brought into the laboratory. No eating, drinking or smoking should be permitted in the laboratories. No mouth operations should be allowed in the laboratory. Labels should be moistened with a wash bottle, not by mouth. Protective gloves and clothing should be worn and these should be monitored for contamination after use. The working surfaces in the laboratory should be of non-absorbent material and the sources should be held in trays as shown in figure 3.2. Where required the sources should be handled in fume cupboards connected directly to the outside atmosphere. The sources should be moved in double containers. For example, the ampoule containing a source can be moved using the container in which it is normally supplied. In the United Kingdom, chemical inspectors will give advice on the preparation of a suitable laboratory. The disposal of waste is very often the main problem with open sources and this can usually be done by disposing up to a maximum activity per month down a sink in liquid form. Solid waste is usually disposed of in conjunction with the local authority refuse service. Records must be kept of waste disposal. All surfaces and equipment in the laboratory should be surveyed with an appropriate contamination monitor. Standards are set for maximum permissible surface contamination for the differing radionuclides in terms of the activity of the material and for the concentration of radioactive materials in water and air and these must be adhered to. These standards can be found in the various Codes of Practice listed in the Bibliography (page 82).

Personnel leaving an open source laboratory should be monitored for contamination.

Methods for decontamination

Should a liquid spill occur, drop a handful of paper tissues on the site of the spill and mop up. Dispose of the tissues as waste or if the source has a short half-life, wait until the activity has decayed. Monitor the surface when dry. If a solid spill occurs, cover the spill with damp tissues and again mop up and monitor the surface. Should there be any contamination, the *Code of Practice for the Protection of Persons Exposed to Ionising Radiations in Teaching and Research* suggests the following methods for decontamination

(1) Spilt liquid should be absorbed on paper tissue or 'Vermiculite'. Where dry material has been spilt and there is loose particulate contamination, the best method of decontamination may well be the application of a strippable coating by brush or spray. This coating will hold the contaminant and prevent the dispersal (provided that the method chosen does not disturb the loose contaminant). When dry, the coating is stripped off taking the adhering contaminant with it. Self-adhesive tapes can also be applied to non-porous surfaces for the removal of loosely held dust.

(2) For the second stage of decontamination, the use of damp swabs is preferable to uncontrolled sluicing to prevent spread of contamination. The actual method to be used and the appropriate agent will depend to some extent on the ease of removal of the remaining contamination, and the methods set out below are arranged in order, to deal with increasingly difficult circumstances

    (a) Treat with a suitable detergent which may be in the form of a cream to prevent the spread of the contamination by splashing. Swabbing or light scrubbing may also be necessary.

(b) Scrub lightly with a complexing solution. If an application of a thickened complexing agent is used, it should be left on the surface for a few hours before rinsing off. The addition of pigment will help to identify the areas to which the decontaminant has been applied.

(c) Swab or scrub with mild abrasive pastes containing complexing agents.

(d) Where none of the above methods is successful and the contamination still remains, it will be necessary to treat the surface by more vigorous scrubbing and abrasion or more severe treatment.

Contaminated clothing should be removed and the following decontamination measures should be taken where appropriate.

Wounds

Contaminated wounds should be washed under a fast-running tap and bleeding should be encouraged. Care should be taken not to contaminate the eyes, mouth and nostrils. The wound should finally be washed with soap and water, treated with mild antiseptic and dressed.

Mouth

The mouth should be washed out several times with hydrogen peroxide solution (1 tablespoon of 10-volume solution to a tumbler of water).

Eyes

Eyes should be irrigated with saline solution (0·9 per cent common salt solution) or else with tap or distilled water.

Hands

Hands should be washed with soap, water and a soft nail brush. If this is unsuccessful, they should be washed with EDTA soap. As a last resort, the hands may be immersed in concentrated permanganate solution, allowed to dry and finally wiped with five per cent sodium bisulphite solution.

Other areas of skin

Other areas of skin should be rubbed gently with cotton wool soaked in 'Cetavlon' taking care not to damage the skin.

General

Decontamination should continue in all cases until monitoring shows that contamination has been reduced to an acceptable level, unless there is a danger of contamination entering the blood stream via the skin roughened or broken by decontamination procedures.

Materials required for decontamination should be kept readily available.

Transport of radioactive materials

The transport of radioactive materials is covered by the *Code of Practice for the Carriage of Radioactive Materials by Road* published by Her Majesty's Stationery Office. All persons who regularly send or carry radioactive materials by road should have a copy of this booklet, as it outlines the procedures to be adopted. These are intended to reduce to a minimum the hazards which may arise in case of accidents and also to ensure the safety of personnel regularly travelling with active materials. The hazards to be minimized as always are (a) the prevention of overexposure by external radiation, and (b) the prevention of the intake of radioactive materials.

Since the increasing use of radioactive materials in industrial processes such as radiography makes it increasingly likely that at some time members of the general public, the police and fire brigade may well be involved in incidents involving radioactive materials, this section is devoted to outlining the procedures which should be followed and to providing information which could be useful in such a situation.

Packaging

There are four types of packaging used in transporting radioactive materials. Two of the types, commercial and industrial packaging, need not cause much concern since in these cases if all the material was released from the packages there would still be no great hazard.

Type A and type B packages are used for small and large quantities of radioactive materials respectively, and in this case if there was leakage from the package a hazard would result. Both of these types of package are designed to specific standards and must pass tests regarding their behaviour when subjected to fire and shock. There are maximum quantities of material which can be shipped in these packages according to the radionuclide being transported. All radionuclides as already stated are not equally hazardous so they are separated into seven transport groups according to their toxicity, and the group a nuclide is classed into sets upper limits to the quantity of material which can be consigned in type A and B packages. The labelling of packages is standardized so that the hazard from any particular package can quickly be assessed.

Represents yellow   Represents red

Figure 9.2   A type B package label

Labelling procedures

The following labelling procedures are adopted

Category I   White label   The dose rate at the surface of the package is 0·5 mrem/h or less

Category II   Yellow label   The dose rate at the surface of the package does not exceed 10 mrem/h and the dose rate 1 m from the centre of the package does not exceed 0·5 mrem/h

Category III   Yellow label   The dose rate does not exceed 200 mrem/h at the surface of the package and 10 mrem/h 1 m from the centre of the package

Thus, if necessary, undamaged white label packages can be handled with no ill effects. Yellow label packages on the other hand, can only be handled for limited times.

A typical label is shown in figure 9.2. Half of this label consists of the standard radioactive warning sign. The other half lists the principal content of the package and its activity. Also given on type B labels is the transport index of the package. This gives the dose rate at 1 m from the centre of the package, when undamaged.

Road accidents

Suppose that a vehicle carrying radioactive material is involved in a road accident. If the driver is unhurt he should know what steps to take. But suppose he is unable to help. What should be done? If the correct procedures are being followed, the first warning that the vehicle is carrying radioactive material will be given by the vehicle labels attached to the side of the vehicle. Additional notices will be found inside the vehicle again giving warning in the case of accident. *Clearly it is important that these notices be displayed and that if the vehicle is not carrying radioactive material they should be removed.* When it is clear that the vehicle is carrying radioactive materials there should be no undue alarm. All the type B packages are designed to withstand the effects of severe accidents and all the type A packages are not likely to present any serious hazard.

Firstly, if possible remove the driver and any injured persons to a safe distance from the vehicle or vehicles involved in the accident.

Wear gloves while doing this if you have them. Position the injured persons behind any shielding such as a stone wall, if readily available. Then notify the police. In the event of damage from fire or where rescue tackle is needed, the fire brigade should also be notified. At this stage keep the public away from the vehicle subject to the overriding need of saving life. If any shielding material is visible inside the vehicle, observe what type it is. Observe the type of radiation instrument which may be in the cab of the vehicle. Do not attempt to remove the load from the vehicle but try to contain any spilt radioactive material. Alpha and beta particles will be stopped by the vehicle body. Always approach the vehicle using the engine as a shield. This will absorb most gamma-rays. If any person is thought to have radioactive material on his body he should not smoke, eat or drink. Any contaminated clothing should be removed as soon as is convenient. See if the driver has a consignment note which will be of use to the police and fire brigade.

The police on arrival can if necessary summon expert advice under the NAIR scheme—National Arrangement for Incidents Involving Radioactivity. Whether such advice is required or not can only really be determined by surveying the vehicle and its surroundings with a dose rate meter of the type previously described and, if contamination is suspected, with a contamination meter. If the police or fire brigade cannot do this and suspect that any of the packages have been damaged then they almost certainly should get expert advice. ·

All people transporting radioactive materials should consider the possible hazards to others and ensure that they obey the rules for transporting radioactive materials.

Things to remember from chapter 9

(1) Before possessing, using, selling or transporting radioactive materials, it is necessary in most countries to register with an appropriate authority.

(2) The regulations which are in force in the United Kingdom place certain obligations on the controlling authorities. For example, it is usually necessary to appoint a competent person to ensure that the regulations are adhered to.

(3) Workers must be classified into designated and non-designated persons.

(4) Designated persons will be subject to special supervision and

should be given special training concerning the hazards involved and precautions to be observed when using radioactive materials.

(5) It is important to follow the *Code of Practice for the Carriage of Radioactive Materials by Road.*

(6) In particular, all vehicles carrying radioactive materials should be clearly labelled and all packages should be up to standards required for radioactive packages.

10

Forewarned is forearmed

Since this book is intended primarily for the layman who encounters radioactive materials in the course of his work, it would not be complete without a brief review of some of the principal ways in which radioactive materials are used in industry. The use of sealed sources in radiography, in thickness and level gauges and in density and moisture content measurement will be considered in turn. Some applications of open sources will also be mentioned.

Radioactive materials are widely used in medicine. These uses are not considered here since they are highly specialized and personnel involved will almost certainly be given training sufficient to meet any situation they encounter. A brief mention is made, however, of the use of x-rays in dentistry and the precautions which should be taken by the dentist.

Radiography

This is one of the most widely used techniques in the non-destructive testing of welds, castings and so on. In the process, gamma- or x-radiation is passed through the object under examination. A film is placed behind the object and a shadow 'photograph' of the object is then obtained when the film is developed. If there is a hole in a weld, or if the object is thinner at one place than another, then more gamma- or x-rays will pass through these regions. There is a blackening of the radiograph corresponding to these regions, and in this way the flaws in castings can be located. The most commonly used gamma sources are listed in table 10.1 together with their half-lives and the energies of the principal gamma-rays they emit.

Table 10.1 Gamma sources used in radiography

Source	Half-life	Gamma-ray energies (MeV)	Γ (rem/h per curie at 1 m)*
Cobalt-60	5·3 y	1·17, 1·33	1·32
Caesium-137	30 y	0·66	0·33
Iridium-192	74 d	0·30 to 0·61	0·48
Thulium-170†	127 d	0·052, 0·084	0·0025

* Γ (the Greek capital letter gamma) is the usual symbol used for the specific gamma-ray constant.

† In this source a small quantity of higher energy *bremsstrahlung* radiation is also present.

Also given is the specific gamma-ray constant for these sources. This is the dose rate/h at 1 m from the source per curie activity. Radiography sources are in general the most hazardous of all sources used in industry. The dose rate is usually very large in any area where a radiography source is exposed so the sources must be shielded when not in use. The sources are therefore kept in 'exposure containers' which are usually made of thick lead although tungsten alloy and depleted uranium are being used to an increasing extent.

Types of container used

Three general types of container are in use. In the shutter type of container, opening the shutter exposes the source. A typical shutter type container is shown in figure 10.1. In the 'torch' container the source is mounted in a removable portion of the shield which can be detached from the main shield and held in the hand like a torch. This 'torch' must never be pointed at anyone. In the third type of container the source is moved to the exposure position by remote control. In some ways this ought to be the safest to use but in fact there have been many incidents arising when the source has not been returned to its container. With this type of container, it is essential to use a dose rate meter to check that the hazard has been removed and that the source has returned to its container. The *British Standards Institution Booklet BS 4097* (1966) gives a clear account of what is required of an exposure container. It should be strong, stable and weatherproof. It should have handles or lugs for lifting and there should be a lock to hold the source or shutter in the safe position. *It should be stressed that containers are usually designed to provide a safe shield for a specific type and activity of source. A container which is designed for*

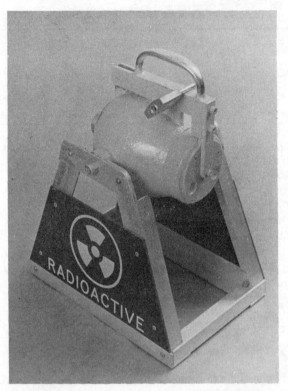

Figure 10.1 A shutter type container for radiography. Note that the shutter is locked when the source is not being used

use with a 10 Ci iridium source is most unlikely to be safe to use with a 10 Ci cobalt source. Greater thicknesses of lead are required to shield a cobalt source than an iridium source of equal activity.

In moving the sources from site to site or in replacing the sources, the *Code of Practice for the Transport of Radioactive Materials* must be adhered to. *In transporting an exposure container, it is wise to place it in a second container which should be fireproof if possible.*

X-ray generators

Where radiography is part of a production process, x-ray generators are usually used. The x-ray generator is housed in a room completely lined with lead or with walls made of thick barium concrete. The thickness of shielding required varies of course with the energy of the x-rays produced. Since the dose rate from x-ray sets is usually very large, arrangements must be made so that the generator can be shut off quickly by personnel who may accidentally be in the radiation area when the generator is switched on. Warning lights or audible signals should be provided to give adequate warning that the generator is about to be used.

Dental x-rays

The use of x-rays in dentistry is a widespread practice and the same precautions are required as in other types of radiography. The use of x-rays is covered in the *Code of Practice for the Protection of Persons against Ionising Radiations arising from Medical and Dental Use* and this booklet should be consulted. In general, the size of the beam of x-rays should be just sufficient to allow the radiographs of the teeth to be taken. Patients should use lead aprons. X-rays should be used as sparingly as possible. In the United Kingdom, the National Radiological Protection Board runs a special service for dentists. They survey the generators when in use and provide a three-month trial personnel monitoring service for the dentist and his staff. It can then be decided if the monitoring is needed or not.

Level, backscatter and thickness gauges

These gauges are fairly widely used in industry. *Level gauges* consist simply of a source of radiation and a detector. When the level of any material rises so that it comes between the source and detector, the count rate or dose rate at the detector decreases. This indicates that the level has been reached. Very often these gauges are used in production lines to detect unfilled packages. The sources most often used are given in table 10.2. The activity of the sources is not usually greater than about 100 mCi so they are not as hazardous as radiography sources. In some cases, the activity of the source may be sufficiently low to permit the issue of a Certificate of Exemption (no. 4) General by H.M. Inspectorate. In these cases, film badges do not have to be used.

Backscatter gauges consist of a beta source and detector. The thickness of very thin layers of material can be determined by measuring the number of beta particles backscattered from the material. When beta particles fall on the surface of materials, some of them are

Table 10.2

Gauge	Type of source
Thickness gauges	(a) For high density materials, x-ray generators giving low energy x-rays are used
	(b) High energy β sources are used for thick sheets of low density material such as roofing felt, lino, rubber. The source is usually strontium-90/yttrium-90
	(c) Medium energy beta souces are used for thin sheets of low density materials such as paper; usually thulium-204 or krypton-85 are used
	(d) For very thick layers, gamma sources are used
Backscatter gauges	Usually beta sources are used; strontium-90/yttrium-90, thulium-204, caesium-137, promethium-147 and carbon-14 are the most commonly used sources.
Level gauges	Gamma-ray sources are used, usually caesium-137 or cobalt-60

reflected back in the direction from which they came—backscattered. The number of particles backscattered increases with the thickness of the material but quickly reaches a maximum value so that only the thickness of thin layers can be measured in this way.

Thicker layers can also be measured using *thickness gauges*. Here the number of β particles or x-rays passing through a layer of the material is measured. This number is directly related to the thickness. For really thick layers, gamma sources are sometimes used. The housing of the gauge should be clearly marked to show that it contains radioactive material.

When these gauges are in use it is often difficult to provide adequate shielding so protection is achieved by the use of distance. Barriers are erected where the dose rate exceeds 0·75 mrem/h to prevent access to the radiation area round the source. With these gauges a shutter should be used if possible to shield the source when it is not in use. A clear indication should be given when the shutter is open or closed.

Moisture and density measuring equipment

Moisture and density gauges are increasingly being used in building and construction, mining, the steel industry, geological surveying and in the soil sciences. The advantage they have is that measurements of density or moisture content can be made very rapidly. In addition the methods are non-destructive.

Density measurements are obtained using a gamma source and a geiger counter. The gamma-rays are allowed to fall on the material and the number of gamma-rays which are scattered towards the detector is used to give the density of the material. This number increases with the density. The most commonly used source is cobalt-60. The activity is usually of the order of 5 mCi so that the dose rate at 1 m is about 6 mrem/h. The usual precautions with gamma sources should therefore be observed.

Moisture content is measured in similar fashion only in this case neutron sources are used. The neutrons emitted from the source are moving rapidly (high energy) but are slowed down by banging into the nucleus of the hydrogen atoms present in the material. Some of these slowed neutrons escape from the material and the number which escape is measured using a neutron detector. This number is determined by the moisture content of the material.

Americium–beryllium sources (Am–Be) are most commonly used. A typical source of 100 mCi emits $2 \cdot 7 \times 10^5$ neutrons/s or 270 000 neutrons/s. The dose rate at 1 m from this source is approximately 1/3 mrem/h. The sources used in these gauges thus do not present any serious external radiation hazard. Sometimes radium–beryllium sources are used for combined density–moisture content measurements. These emit both gamma-rays and neutrons. With these sources, the gamma-rays present the more serious hazard. From a 100 mCi radium–beryllium source the neutron dose rate is approximately $1 \cdot 5$ mrem/h at 1 m, but the gamma dose rate is about 80 mrem/h.

Other uses of gamma sources

Gamma sources are used in other ways from those already mentioned. They are used for sterilizing instruments, for producing genetic changes in seeds and in various chemical uses. Cobalt sources are used in the manufacture of car tyres. Here the carbon atoms which form the main part of the tyre are bonded together directly using the gamma-radiation. Identification and measurement of the composition of materials can be made by measuring the characteristic x-rays emitted when the materials are exposed to gamma- or x-radiation. This is called x-ray fluorescence spectrometry.

Dispersal of static electricity

Most people will have observed the effects of static electricity. For example, on boarding a bus you may at some time have experienced a slight electric shock on gripping the mounting rail. This is caused by static electricity passing through you to earth. This static electricity is liable to occur on any insulating material and in many processes it constitutes a fire hazard. The static electricity can be dispersed by using an alpha or beta source to ionize the air surrounding the material. The air becomes conducting, the static electricity then leaks away and the fire risk is reduced. Radioactive materials are used for this purpose in the manufacture of paper, rubber, or textiles, in photography and in handling liquids like petrol and benzene. Usually a strontium-90/yttrium-90 beta source is used.

Sometimes alpha sources are used. Figure 10.2 shows a picture of a static master brush which is used in photography for brushing films and cleaning lenses. This incorporates a polonium-210 source which emits alpha particles. The polonium is sealed in small glass beads. The beads are glued to a strip of material so that the source resembles the side of a matchbox. Since the alpha particles are stopped by a few centimetres of air they present no great hazard unless the source is

Figure 10.2   A static master brush. These are no longer being imported into the United Kingdom

held close to the eye. The main danger is that some of the small beads might become unstuck. In the United Kingdom, if private individuals purchase a static master they are still required to register with the appropriate authority. Rather surprisingly this is not the case in the United States of America, but in the United Kingdom the importing of these brushes has recently been stopped.

The use of unsealed sources in industry

So far only the uses of sealed sources have been considered. Open source material is also used. A typical use is in the manufacture of radio valves. Radium and tritium are often incorporated in the valves. When handling tritium, it should be remembered that tritium in the form of tritiated water can be absorbed directly through the skin. If the maximum permissible body burden is exceeded it will probably be no consolation that the emergency procedure is to drink as much beer as possible to make oneself sick. When the manufactured valves are stored, the number stored in any one place must be limited because of the external hazard they will then present.

Radioactive tracers are also widely used. Here the spread of the radioactive material can be used to indicate how well some process is working. Another use is in the manufacture of luminous watches or luminous materials. At one time radium was used in this way, but nowadays the use of tritium or krypton is more common. These low-energy β emitters are mixed with zinc sulphide and when the beta particles are stopped by the material, light is given off—hence the luminous appearance. Powerful signal lamps requiring no source of power can be made in this way. The low-energy β emitters are used in preference to the radium, because the β particles are more readily stopped by the glass of the watch face than the gamma-rays from radium.

A final word of advice

Remember that the safety of yourself and your fellow workers depends on you. Always follow the general principles of radiation protection outlined in this book. If this is done, then radioactive materials are probably less hazardous than many other toxic materials used in industrial processes. *It is only when the rules are ignored that*

*there is any real danger. If you ignore the rules, you may put yourself
and others at risk.*

Things to remember from chapter 10

(1) Gamma sources are used in radiography.
(2) The gamma sources are kept in exposure containers. When using a remotely controlled container, it is important to check, using a suitable dose rate meter, that after the source has been returned to its container.
(3) The thickness of the container must be chosen to suit the type and activity of the source.
(4) Level gauges usually use gamma-ray sources.
(5) Thickness and backscatter gauges use beta sources.
(6) Density measurements are made using gamma sources.
(7) Moisture content measurements are made using neutron sources.
(8) Combined moisture–density gauges usually use radium–beryllium sources which emit both neutrons and gamma-rays.
(9) Alpha and beta sources are used for the dispersal of static electricity.
(10) Static master brushes contain an alpha source and must be registered even if used by private individuals.
(11) Open source material is also used in industry. Normal precautions for open source materials must always be observed.
(12) Remember your safety and the safety of your fellow workers depends on the care you take.

Appendix—Classification tables of radionuclides

The Medical Research Council has adopted the Classification of Radionuclides according to toxicity proposed by the I.A.E.A. in 1963 (Technical Reports Series No. 15). The list is given in the table below.

Class I Radionuclides (high toxicity)

Radionuclide	Symbol	Mass number
Lead	Pb	210
Polonium	Po	210
Radium	Ra	223
Radium	Ra	226
Radium	Ra	228
Actinium	Ac	227
Thorium	Th	227
Thorium	Th	228
Thorium	Th	230
Protoactinium	Pa	231
Uranium	U	230
Uranium	U	232
Uranium	U	233
Uranium	U	234
Neptunium	Np	237
Plutonium	Pu	238
Plutonium	Pu	239
Plutonium	Pu	240
Plutonium	Pu	241
Plutonium	Pu	242
Americium	Am	241
Americium	Am	243
Curium	Cm	242
Curium	Cm	243

Radionuclide	Symbol	Mass number
Curium	Cm	244
Curium	Cm	245
Curium	Cm	246
Californium	Cf	249
Californium	Cf	250
Californium	Cf	252

Class II Radionuclides (medium toxicity—upper sub-group A)

Radionuclide	Symbol	Mass number
Sodium	Na	22
Chlorine	Cl	36
Calcium	Ca	45
Scandium	Sc	46
Manganese	Mn	54
Cobalt	Co	60
Strontium	Sr	89
Strontium	Sr	90
Yttrium	Y	91
Zirconium	Zr	95
Ruthenium	Ru	106
Silver	Ag	110m
Cadmium	Cd	115m
Indium	In	114m
Antimony	Sb	124
Antimony	Sb	125
Tellurium	Te	127m
Tellurium	Te	129m
Iodine	I	126
Iodine	I	131
Iodine	I	133
Caesium	Cs	134
Caesium	Cs	137
Barium	Ba	140
Cerium	Ce	144
Europium	Eu	152 (13 years)
Europium	Eu	154
Terbium	Tb	160
Thulium	Tm	170
Hafnium	Hf	181
Tantallum	Ta	182
Iridium	Ir	192
Thallium	Tl	204
Bismuth	Bi	207
Bismuth	Bi	210
Astatine	At	211

Radionuclide	Symbol	Mass number
Lead	Pb	212
Radium	Ra	224
Actinium	Ac	228
Protoactinium	Pa	230
Thorium	Th	234
Uranium	U	236
Berkelium	Bk	249

Class III Radionuclides (medium toxicity—lower sub-group B)

Radionuclide	Symbol	Mass number
Beryllium	Be	7
Carbon	C	14
Fluorine	F	18
Sodium	Na	24
Chlorine	Cl	38
Silicon	Si	31
Phosphorus	P	32
Sulphur	S	35
Argon	A	41
Potassium	K	42
Calcium	Ca	47
Scandium	Sc	47
Scandium	Sc	48
Vanadium	V	48
Chromium	Cr	51
Manganese	Mn	52
Manganese	Mn	56
Iron	Fe	55
Iron	Fe	59
Cobalt	Co	57
Cobalt	Co	58
Nickel	Ni	63
Nickel	Ni	65
Copper	Cu	64
Zinc	Zn	65
Zinc	Zn	69m
Gallium	Ga	72
Arsenic	As	73
Arsenic	As	74
Arsenic	As	76
Arsenic	As	77
Selenium	Se	75
Bromine	Br	82
Krypton	Kr	85m
Krypton	Kr	87

Radionuclide	Symbol	Mass number
Rubidium	Rb	86
Strontium	Sr	85
Strontium	Sr	91
Strontium	Sr	92
Yttrium	Y	90
Yttrium	Y	92
Yttrium	Y	93
Zirconium	Zr	97
Niobium	Nb	93m
Niobium	Nb	95
Molybdenum	Mo	99
Technetium	Tc	96
Technetium	Tc	97m
Technetium	Tc	97
Technetium	Tc	99
Ruthenium	Ru	97
Ruthenium	Ru	103
Ruthenium	Ru	105
Rhodium	Rh	105
Palladium	Pd	103
Palladium	Pd	109
Silver	Ag	105
Silver	Ag	111
Cadmium	Cd	109
Cadmium	Cd	115
Indium	In	115m
Tin	Sn	113
Tin	Sn	125
Antimony	Sb	122
Tellurium	Te	125m
Tellurium	Te	127
Tellurium	Te	129
Tellurium	Te	131m
Tellurium	Te	132
Iodine	I	132
Iodine	I	134
Iodine	I	135
Xenon	Xe	135
Caesium	Cs	131
Caesium	Cs	136
Barium	Ba	131
Lanthanum	La	140
Cerium	Ce	141
Cerium	Ce	143
Praseodymium	Pr	142
Praseodymium	Pr	143
Neodymium	Nd	147
Neodymium	Nd	149

Radionuclide	Symbol	Mass number
Promethium	Pm	147
Promethium	Pm	149
Samarium	Sm	151
Samarium	Sm	153
Europium	Eu	152
		(9·2 hours)
Europium	Eu	155
Gadolinium	Gd	153
Gadolinium	Gd	159
Dysprosium	Dy	165
Dysprosium	Dy	166
Holmium	Ho	166
Erbium	Er	169
Erbium	Er	171
Thulium	Tm	171
Ytterbium	Yb	175
Lutecium	Lu	177
Tungsten	W	181
Tungsten	W	185
Tungsten	W	187
Rhenium	Re	183
Rhenium	Re	186
Rhenium	Re	188
Osmium	Os	185
Osmium	Os	191
Osmium	Os	193
Iridium	Ir	190
Iridium	Ir	194
Platinum	Pt	191
Platinum	Pt	193
Platinum	Pt	197
Gold	Au	196
Gold	Au	198
Gold	Au	199
Mercury	Hg	197
Mercury	Hg	197m
Mercury	Hg	203
Thallium	Tl	200
Thallium	Tl	201
Thallium	Tl	202
Lead	Pb	203
Bismuth	Bi	206
Bismuth	Bi	212
Radon	Rn	220
Radon	Rn	222
Thorium	Th	231
Protoactinium	Pa	233
Neptunium	Np	239

Class IV Radionuclides (low toxicity)

Radionuclide	Symbol	Mass number
Hydrogen	H	3
Argon	A	37
Cobalt	Co	58m
Nickel	Ni	59
Zinc	Zn	69
Germanium	Ge	71
Krypton	Kr	85
Strontium	Sr	85m
Rubidium	Rb	87
Yttrium	Y	91m
Zirconium	Zr	93
Niobium	Nb	97
Technetium	Tc	96m
Technetium	Tc	99m
Rhodium	Rh	103m
Indium	In	113m
Indium	In	115
Iodine	I	129
Xenon	Xe	131m
Xenon	Xe	133
Caesium	Cs	134m
Caesium	Cs	135
Samarium	Sm	147
Rhenium	Re	187
Osmium	Os	191m
Platinum	Pt	193m
Platinum	Pt	197m
Thorium	Th	232
Natural thorium	Th-Nat	
Uranium	U	235
Uranium	U	238
Natural uranium	U-Nat	

Note 1: The I.A.E.A. classification does not list seven radionuclides which are being used in the U.K. These are:

Radionuclide	Class
Oxygen O-15	IV
Potassium K-43	III
Iron Fe-52	III
Cobalt Co-56	II
Iodine I-124, I-125	II
Iodine I-130	III

Note 2: 'm' indicates metastable state. This is the name given to excited nuclear states which take longer to decay to their ground state than other states of the same type. The term is used to identify the half-life to use with the nuclide.

Bibliography

This is not a complete bibliography, as it only includes the publications referred to in the book and a few others which may be useful.

Copies of the Acts, Regulations and Codes of Practice listed in this Bibliography are available from H.M. Stationery Office except when stated.

Relevant acts and regulations in the United Kingdom

Radioactive Substances Act (1960)

This Act regulates the keeping and use of radioactive material and makes provision as to the disposal and accumulation of radioactive waste.

Regulations under the Factories Act (1961)

(a) The Ionising Radiations (Sealed Sources) Regulations (1969)
(b) The Ionising Radiations (Unsealed Radioactive Substances) Regulations (1968)

These regulations impose requirements for the protection of persons employed in factories and other places to which the Factories Act (1961) applies.

Regulations affecting the transport of radioactive materials

(a) The Air Navigation Order (1970)
(b) The Merchant Shipping (Dangerous Goods) Rules (1965)

(c) Inland Postal Regulations
(d) British Railways Conditions of Carriage for the Conveyance of Radioactive Substances by Merchandise and Passenger Train
(e) The Radioactive Substances (Carriage by Road Great Britain) Regulations (1974)
(f) The Radioactive Substances (Road Transport Workers) Regulations (1970)

Important non-statutory codes of practice

(a) Code of Practice for the Protection of Persons against Ionising Radiations arising from Medical and Dental Use
(b) Code of Practice for the Protection of Persons exposed to Ionising Radiations in Research and Teaching
(c) Memorandum on the Use of Ionising Radiations in Schools, Establishments of Further Education and Teacher Training Colleges
(d) Code of Practice for the Carriage of Radioactive Materials by Road
(e) Code of Practice for the Storage of Radioactive Materials in Transit

International codes

(a) The Manual of Industrial Radiation Protection: Part II. Model Code of Safety Regulations (Ionising Radiations) (1959)
(b) International Atomic Energy Agency Regulations for the Safe Transport of Radioactive Materials (1967)

A survey of the existing legislation for different countries has been published by the World Health Organization in their *International Digest of Health Legislation* and off-prints of the relevant parts (Protection against Ionising Radiations) may be obtained from

> WHO
> Distribution and Sales Unit,
> Palais des Nations,
> Geneva,
> Switzerland

Useful British Standards Institution Booklets

British Standards Institution Booklet BS 4097 (1966)
British Standards Institution Booklet BS 4094, Part 1 (1966)

International Atomic Agency, Vienna: Safety Series

This series covers a wide range of topics.

International Commission on Radiological Protection Publications

These are available from Pergamon Press.

Index

Date Due

Due	Returned	Due	Returned
FEB 2 1 '78	FEB 0 8 '78		
FEB 2 6 '78 NOV 1 3 '78	MAR 1 5 '78		
DEC 8 '78	NOV 2 5 '78 NOV 1 2 '80		
JUN 8 1981	JUN 5 '81		
AUG 7 '81	JUL 9 '81		
DEC 1 8 '81	DEC 2 1 '81		
MAY 1 4 1982	APR 2 6 1982		
JUN 1 1 1982	JUN 1 1 1982		
NOV 0 7 1982	NOV 0 7 1982		
NOV 2 8 1982	NOV 2 8 1982		
MAR 2 4 1983			
MAR 2 4 1983	MAR 2 4 1983		
DEC 0 8 1985 DEC 1 1 1986	DEC 0 5 1985 DEC 0 1 1986		
APR 2 7 1992	4 1992		
APR 1 1 2002	APR 1 5 1992 0 2 2002		
AUG 0 7 2002	AUG 0 2 2002		